Inclusive Intelligence

AI for People with Disabilities

By
Maya Jordan

Copyright 2024 Lars Meiertoberens. All rights reserved.

No part of this book may be reproduced in any form or by any electronic or mechanical means, including information storage and retrieval systems, without permission in writing from the author. The only exception is a reviewer who may quote brief excerpts in a review.

Although the author and publisher have made every effort to ensure that the information in this book was correct at the time of going to press, the author and publisher accept no liability to any party for any loss, damage or disruption caused by errors or omissions, whether such errors or omissions are due to negligence. accident or any other cause.

This publication is intended to provide accurate and reliable information with respect to the subject matter covered. It is sold on the understanding that the publisher does not provide professional services. If legal advice or other expert assistance is required, the services of a competent professional should be sought.

The fact that an organization or website is mentioned in this work as a citation and/or potential source of further information does not imply that the author or publisher endorses the information the organization or website provides or the recommendations it makes.

Please keep in mind that websites listed in this work may have changed or disappeared between the time this work was written and the time it was read.

Inclusive Intelligence

AI for People with Disabilities

Table of Contents

Introduction 1

Chapter 1: Understanding Disabilities 5
    Overview of Different Disabilities 5
    Common Challenges and Barriers 8

Chapter 2: Introduction to Artificial Intelligence 13
    Basic Concepts of AI 13
    History and Evolution of AI 17

Chapter 3: The Intersection of AI and Disability 21
    The Role of AI in Accessibility 21
    Potential Benefits and Risks 24

Chapter 4: AI-Powered Assistive Technologies 28
    Overview of Assistive Tech 28
    Case Studies and Examples 32

Chapter 5: Smart Assistants and Personal Aides 36
    Virtual Assistants for Daily Living 36
    Enhancing Communication and Interaction 40

Chapter 6: Vision and Hearing Impairment Solutions 45
    AI for the Visually Impaired 45
    AI for the Hearing Impaired 48

Chapter 7: Mobility and Physical Disabilities 52
    AI in Mobility Solutions 52
    Robotic Assistance and Adaptability 56

Chapter 8: Cognitive Disabilities and AI 61
  AI Tools for Cognitive Support.. 61
  Improving Learning and Memory 66
Chapter 9: AI in Education 70
  Personalized Learning Environments 70
  Bridging Educational Gaps.. 74
Chapter 10: Healthcare Innovations 79
  AI in Medical Diagnostics 79
  AI for Personalized Healthcare 82
Chapter 11: Mental Health and Emotional Support 86
  AI for Mental Health Monitoring 86
  Emotional Wellbeing Tools 90
Chapter 12: Enhancing Workplace Accessibility 94
  AI in the Workplace... 94
  Tools for Remote Work and Collaboration................................. 98
Chapter 13: AI in Communication Aid 102
  Advanced Communication Devices 102
  Natural Language Processing Applications.............................. 106
Chapter 14: Legal and Ethical Considerations 111
  Privacy Concerns 111
  Fair and Inclusive AI Development 115
Chapter 15: Community and Social Inclusion................................ 119
  AI for Community Engagement.. 119
  Enhancing Social Interaction 123
Chapter 16: User-Centered Design 128
  principles of inclusive design 128
  Involving Users in Development... 132
Chapter 17: Challenges and Barriers in Implementation 136
  Technical and Societal Challenges 136

Addressing Accessibility in AI Systems..139
Chapter 18: Future Trends and Innovations......................................144
    Emerging AI Technologies..144
    Predicted Developments...148
Chapter 19: Global Perspectives..152
    Case Studies from Around the World..152
    International Standards and Practices..157
Chapter 20: Collaboration Between Stakeholders...........................162
    Partnerships in Development...162
    Roles of Government and NGOs..166
Chapter 21: Funding and Support for AI Projects...........................170
    Sources of Funding..170
    Grants and Incentives..174
Chapter 22: Accessibility Standards and Guidelines........................178
    Current Standards..178
    Implementing Best Practices..182
Chapter 23: Case Studies of Successful AI Applications.................186
    Detailed Examples...186
    Lessons Learned...191
Chapter 24: Building an Inclusive AI Ecosystem..............................195
    Strategies for Inclusivity..195
    Community Involvement...199
Chapter 25: Personal Stories and Testimonials.................................203
    Real-Life Experiences..203
    Impact of AI on Individual Lives..207
Conclusion..212
Appendix A: Appendix..216
    Supplementary Resources...216

Glossary of Terms 217
Case Study Summaries 217
Toolkits and Checklists 217
Contact Information 217
Additional Case Studies and Testimonials 218
Research and Development Projects 218

Glossary of Terms 219
    Additional Resources 221

Introduction

The power of artificial intelligence (AI) stretches far beyond mere technological marvels—it's a beacon of hope, especially for those living with disabilities. In a world where inclusivity isn't just a goal but an imperative, AI emerges as a transformative agent that can bridge innumerable gaps, making life not just livable but vibrant and fulfilled.

The vast universe of disabilities encompasses a wide array of conditions that impact individuals in myriad ways. These can range from difficulties in mobility and sensory impairments to cognitive challenges and communication barriers. Each disability presents unique hurdles that require specialized approaches. Yet, despite the diversity of challenges, AI offers versatile solutions that can be tailored to meet individual needs. This book aims to shed light on these innovative technologies and practical applications, showing how AI stands as a pillar of support for people with disabilities.

For educators, healthcare professionals, tech enthusiasts, and individuals with disabilities, understanding how AI can be leveraged to enhance accessibility and independence could change lives. Imagine classrooms where learning is personalized for neurodivergent students, virtual assistants that become an indispensable part of daily routines, or medical diagnostics that preemptively identify health issues before they become critical. These are not far-off dreams but tangible realities being shaped today.

The potential of AI in assisting those with disabilities isn't just technical—it's deeply human. It's about giving a voice to the voiceless, legs to those who can't walk, and vision to the blind. It's about ensuring that everyone, regardless of their physical or cognitive abilities, can interact with the world around them on their terms. The stories and examples provided within the chapters of this book serve as a testament to the profound impact AI can have when applied thoughtfully and inclusively.

Though technology is often seen as cold and impersonal, the application of AI in assisting those with disabilities brings warmth and compassion into focus. Consider the impact of a smart assistant that can remind someone with memory challenges to take their medication, or an AI-driven app that can translate sign language into spoken words, thereby breaking communication barriers. These aren't just devices but companions that enhance the quality of life for so many.

Of course, the journey of AI's integration into the realm of disabilities isn't without its challenges. There are ethical considerations, technical barriers, and societal hurdles that must be navigated skillfully. This book will address these issues head-on, offering insights into how best practices and inclusive design can create technologies that truly serve their intended purpose. Moreover, by involving users in the development process, the end products become not only more effective but also more cherished and trusted by those who depend on them.

AI's foray into disability support is a testament to human ingenuity and compassion. It's a collaborative effort that pulls in expertise from various fields—technologists, designers, healthcare workers, and the disability community itself. This collective endeavor ensures that the solutions developed are not just innovative but also empathetic and user-centric.

As we navigate through the pages of this book together, you'll be introduced to groundbreaking projects and real-world applications that highlight the transformative power of AI. From cognitive support tools and advanced communication devices to healthcare innovations and workplace accessibility enhancements, the spectrum of AI's application is vast and inspiring. Each chapter unfolds a new dimension of how AI can uplift lives, making everyday activities more accessible, thus fostering a more inclusive society.

It's fundamental to recognize that AI isn't a one-size-fits-all solution. The beauty of AI lies in its adaptability and personalization. By real-time learning from interactions and experiences, AI systems can tailor their functionalities to meet individual needs, ensuring that the support provided is not just appropriate but also evolves as those needs change over time.

This book delves deep into the symbiotic relationship between AI and disability support, exploring how technology can be a driving force in shaping an inclusive future. The journey isn't merely about technological advancement but about making a tangible difference in people's lives. Every successful implementation of AI in this domain is a step toward a world where everyone, irrespective of their disabilities, can experience equal opportunities, independence, and a higher quality of life.

The pursuit of inclusivity through AI is not just a technological or scientific challenge—it's a moral imperative. It calls on all of us to reimagine and reshape our world so that it becomes a tapestry of varied abilities, each thread as vibrant and vital as the next. Through concerted efforts and collaborative spirit, the dream of an inclusive society powered by AI is well within our reach.

We stand at the precipice of a new era—an era where artificial intelligence isn't merely a tool, but a partner in our journey toward a more inclusive world. This book is your guide to understanding, embracing, and championing this transformative journey. Welcome to an exploration of how AI can and is transforming lives, empowering individuals, and creating a future where everyone has their place, and everyone belongs. It's an inspiring odyssey that showcases human resilience, ingenuity, and the unyielding commitment to making the world better for all.

Chapter 1:
Understanding Disabilities

Understanding disabilities forms the cornerstone of genuinely enhancing accessibility and promoting equality through technology. Disabilities encompass a wide range of physical, cognitive, sensory, and emotional conditions, each presenting unique challenges and requiring tailored solutions. Whether considering the barriers faced by those with mobility issues or the intricacies of supporting cognitive impairments, the landscape is diverse yet interconnected in the shared goal of overcoming limitations. Recognizing these challenges is the first step toward harnessing the power of AI to create transformative, life-enriching tools. By grasping the foundational aspects of disabilities, we can better appreciate the potential of AI-driven innovations to empower individuals and foster a more inclusive society.

Overview of Different Disabilities

Understanding disabilities in their various forms is crucial to finding ways artificial intelligence (AI) can be leveraged to enhance accessibility and improve lives. Disabilities are diverse and can affect individuals in myriad ways, influencing their abilities to perform daily tasks, communicate, move, or even think and learn. Let's delve into the most common types of disabilities and their unique characteristics.

Physical Disabilities: Physical disabilities encompass a range of conditions that impair an individual's ability to move or control their body parts effectively. Some people may be born with congenital disabilities, such as cerebral palsy or spina bifida, while others may acquire disabilities through accidents, illnesses, or conditions like multiple sclerosis or muscular dystrophy. These disabilities often require mobility aids such as wheelchairs, walkers, or prosthetics to assist in daily activities and improve independence.

AI can play a transformative role in this area by introducing innovations such as robotic exoskeletons that aid in mobility or intelligent prosthetics that can mimic natural movements. These technologies not only restore functionality but also enhance the quality of life.

Vision Impairments: Vision impairments range from partial sight loss to complete blindness. Conditions like glaucoma, cataracts, macular degeneration, and diabetic retinopathy are common causes. Some vision impairments can be congenital, while others develop with age or as a result of other health issues. Individuals with vision impairments often rely on assistive technologies like screen readers, braille displays, and magnification software to interact with the digital world.

AI's potential here is enormous, with tools like real-time image recognition and text-to-speech technologies breaking down barriers. For example, AI-driven apps can describe surroundings, read text aloud, or even assist with navigation, enabling people with vision impairments to gain greater autonomy.

Hearing Impairments: Hearing impairments can range from mild hearing loss to profound deafness. Congenital factors, noise exposure, aging, or illnesses can contribute to hearing issues. People with hearing impairments may use hearing aids, cochlear implants, or rely on sign language and captioning to communicate effectively.

One of the promising areas where AI can provide support is through real-time transcription services and advanced hearing aids that adjust to various sound environments. Virtual assistants with improved lip-reading capabilities or sign language recognition could also bridge communication gaps more effectively.

Cognitive Disabilities: Cognitive disabilities encompass a broad spectrum of conditions affecting mental processes like memory, problem-solving, attention, and understanding. These include intellectual disabilities, autism spectrum disorder, dyslexia, and attention deficit hyperactivity disorder (ADHD). People with cognitive disabilities may encounter challenges in learning, communicating, or performing everyday tasks independently.

AI-driven tools can offer personalized learning experiences, adaptive communication aids, and cognitive-behavioral support. AI can analyze patterns in behavior and learning to tailor educational content, aiding in better retention and comprehension.

Mental Health Conditions: Mental health conditions, such as depression, anxiety, bipolar disorder, and schizophrenia, affect a person's emotional well-being and can significantly impair their quality of life. These conditions may be episodic or chronic, and each individual's experience can differ widely.

AI offers significant potential in supporting mental health through tools like chatbots that provide support and monitor mood changes, predictive algorithms that identify early warning signs of mental health crises, and platforms for facilitating access to mental health resources.

Developmental Disabilities: Developmental disabilities are a diverse group of chronic conditions that appear early in life, affecting physical, cognitive, and behavioral development. Conditions such as Down syndrome, autism spectrum disorder, and developmental delays can impact an individual's ability to perform activities essential for daily living.

AI's role here could include early diagnosis through pattern recognition and machine learning, as well as the creation of personalized communication and learning tools that adapt to the developmental stages and needs of the individual.

Speech and Language Disorders: Speech and language disorders can range from difficulty in articulating words (speech disorders) to problems with comprehension and expression of language (language disorders). Conditions such as aphasia, stuttering, and dyspraxia fall into this category.

AI-powered speech recognition software, language processing algorithms, and communication apps hold promise. These tools can assist with voice synthesis for individuals who are non-verbal and improve the efficiency and accuracy of speech therapy.

Each of these disabilities presents unique challenges, but they all share a common need for better accessibility solutions. Utilizing AI, we can design smarter, more adaptive tools that not only compensate for these limitations but also empower individuals to live more independently and confidently. By understanding the wide array of disabilities, we can better appreciate the multifaceted role that AI can play in crafting a more inclusive world.

Common Challenges and Barriers

Understanding disabilities involves recognizing numerous common challenges and barriers faced by individuals. These obstacles can impede daily activities, limit access to opportunities, and affect overall quality of life. However, comprehending these challenges is the first step toward implementing effective AI solutions that can significantly enhance accessibility and independence.

One of the foremost challenges lies in the physical environment. Many public and private spaces are not designed with accessibility in mind. This can include anything from lack of wheelchair ramps and elevators to inadequate signage for the visually impaired. Individuals with mobility impairments, for instance, often encounter difficulties navigating buildings, transportation systems, and even sidewalks. These barriers limit their ability to participate fully in society.

Communication barriers also present significant hurdles. For those with hearing, speech, or cognitive disabilities, effectively communicating their needs or understanding others can be a daily struggle. This issue is compounded by the lack of widespread availability of assistive communication technologies. Sign language interpretation, speech-to-text services, and cognitive aids are not universally accessible, which can isolate individuals from their communities.

Moreover, the digital divide exacerbates these challenges. Technologies designed for the general population often overlook the specific needs of people with disabilities. Many software applications, websites, and digital devices are not user-friendly for those with visual, auditory, or cognitive impairments. The lack of inclusive design principles means that people with disabilities can be left out of digital advancements, which are increasingly essential in modern life.

Educational barriers further complicate the lives of individuals with disabilities. Traditional educational environments and methods often do not accommodate various learning needs. For example, students with dyslexia or attention deficit disorders may struggle with standard teaching techniques, while those with hearing or visual impairments might find classroom settings challenging. The lack of personalized learning tools that cater to diverse abilities limits academic achievement and future career opportunities.

Additionally, employment barriers remain a significant concern. Even as modern workplaces become more inclusive, many people with disabilities still face discrimination and lack of accommodation. The physical setup of workplaces, the inaccessibility of digital tools, and the social stigmas attached to disabilities create substantial obstacles in securing and maintaining employment. These barriers lead to higher unemployment and underemployment rates among individuals with disabilities.

Healthcare disparities are another critical challenge. Access to medical services for people with disabilities is often fraught with obstacles, including inaccessible healthcare facilities, lack of specialized medical equipment, and healthcare providers' insufficient training in disability-specific care. These factors contribute to the overall poorer health outcomes observed among disabled individuals.

Legal and bureaucratic barriers also pose significant challenges. Navigating the legal landscape to secure appropriate accommodations, benefits, or protections can be a daunting task. People with disabilities often face a maze of paperwork, stringent eligibility criteria, and cumbersome processes, which can discourage them from seeking the help they need. Inadequate enforcement of disability rights laws further complicates these issues.

Social attitudes and stigmas around disabilities create additional barriers. Misconceptions, prejudices, and a lack of awareness can lead to social exclusion and discrimination. These societal attitudes can undermine the confidence and self-esteem of individuals with disabilities, causing them to withdraw from social interactions and opportunities.

Economic challenges are also prevalent among this demographic. The additional costs related to managing disabilities, such as healthcare expenses, adaptive equipment, and accessible housing, can be burdensome. People with disabilities often face financial strain, which is exacerbated by limited income opportunities and the high unemployment rate among this group.

Accessibility in the digital realm must also address the diverse and complex needs of users. Websites and applications often lack alternative text for images, keyboard navigation, screen reader compatibility, and other accommodations. This exclusion from digital content can hinder education, employment, and social engagement, making it crucial for AI-driven solutions to prioritize digital inclusion.

Equally important is the challenge of ensuring privacy and security in AI solutions tailored for disabilities. Individuals with disabilities might be more vulnerable to privacy breaches, especially when AI solutions require extensive data collection for personalizing services. Establishing robust data protection measures and maintaining user trust are imperative to addressing this issue.

Training and awareness among developers and stakeholders also present challenges. Many designers and engineers lack sufficient training in inclusive design practices. This gap can lead to unintentional exclusion of disabled users in technological innovations. Raising awareness and integrating disability considerations into AI development processes are essential for creating truly inclusive technologies.

Despite these numerous challenges, the role of advocacy and community support cannot be overlooked. Advocacy groups play a vital role in pushing for policy changes and raising awareness about the needs of individuals with disabilities. Community support systems can provide emotional and practical assistance, facilitating better integration and participation in society.

In concluding this section, it's imperative to recognize that while the barriers are significant, they are not insurmountable. Each challenge presents an opportunity for innovation and improvement.

By harnessing the power of AI and fostering a more inclusive approach, we can create a world that empowers individuals with disabilities, enhancing their quality of life and enabling them to achieve their full potential.

Chapter 2:
Introduction to Artificial Intelligence

Artificial Intelligence (AI) has rapidly evolved from a sci-fi concept to a transformative force in contemporary society, offering considerable promise for people with disabilities. Understanding AI starts with grasping its fundamental principles and how it's reshaping various fields. Primarily, AI involves machine learning, natural language processing, and computer vision, among other technologies, aimed at mimicking human intelligence. Its history dates back to the 1950s, a time when early researchers laid the groundwork for current advancements. Today, AI's continuous evolution presents a unique opportunity to break down barriers and create more inclusive environments. By enhancing everything from accessibility tools to personal aides, AI is not just about innovation; it's about empowerment and improved quality of life for those with disabilities. This chapter aims to lay the foundation for comprehending how AI can redefine opportunities, independence, and overall well-being for individuals with diverse needs.

Basic Concepts of AI

Artificial Intelligence (AI) is a multidisciplinary field that integrates computer science, mathematics, cognitive science, and engineering to create systems capable of performing tasks that typically require human intelligence. At its core, AI aims to develop algorithms and technologies that enable machines to mimic, augment, or even outperform human cognitive functions. These capabilities can range from simple tasks like recognizing patterns to more complex activities like decision-making and problem-solving.

AI can be broadly categorized into two main types: Narrow AI and General AI. Narrow AI, also known as weak AI, is designed to perform a specific task, such as voice recognition, internet searches, or self-driving car navigation. This type of AI operates under a limited scope and doesn't possess the ability to generalize knowledge beyond its predefined tasks. On the other hand, General AI, or strong AI, refers to systems that exhibit human-like intelligence across a wide variety of contexts and can understand, learn, and apply knowledge in a generalized manner.

One of the fundamental concepts in AI is *machine learning*. Machine learning is a subset of AI that focuses on developing algorithms that allow computers to learn from and make decisions based on data. Instead of relying on explicit programming, machine learning systems improve their performance over time by recognizing patterns, making predictions, and adapting to new information. This self-improvement mechanism is particularly impactful for assisting people with disabilities, as it allows for the creation of personalized and adaptive technologies.

Deep learning, a specialized form of machine learning, employs artificial neural networks to model high-level abstractions in data. Inspired by the structure and function of the human brain, these neural networks consist of layers of interconnected nodes (neurons) that process input data to produce outputs. Deep learning has revolutionized fields such as image and speech recognition, making it possible to develop highly accurate assistive technologies for visual and hearing impairments.

To better understand these concepts, consider **natural language processing (NLP)**. NLP is an area of AI that deals with the interaction between computers and humans through natural language. This involves enabling computers to understand, interpret, and generate human language in a way that is both meaningful and useful. Tasks within NLP include language translation, sentiment analysis, and text summarization, all of which can significantly enhance communication for individuals with disabilities.

Another key concept is **computer vision**. Computer vision enables machines to interpret and understand the visual world through the analysis of digital images and videos. Leveraging techniques such as image classification, object detection, and facial recognition, computer vision can provide invaluable support to those with visual impairments by translating visual data into auditory or tactile information.

AI also encompasses **robotics**, a field that integrates AI with mechanical engineering to create machines capable of performing tasks autonomously or semi-autonomously. Robotics plays a crucial role in mobility solutions for individuals with physical disabilities. By combining sensors, processors, and actuators, robots can assist with movements, provide physical support, and even carry out daily chores.

Prediction and decision-making are other critical aspects of AI. *Predictive analytics* uses historical data and statistical algorithms to identify the likelihood of future outcomes. This can be particularly useful in healthcare, where AI can predict potential medical conditions or complications, enabling timely interventions and personalized treatment plans for individuals with disabilities.

In essence, AI aims to enhance human capabilities and improve quality of life by creating systems that can learn, adapt, and function autonomously. While the technical aspects of AI might seem complex, the ultimate goal is simple: to empower individuals by providing intelligent solutions that address real-world challenges. As AI continues to evolve, its potential to transform accessibility and independence for those with disabilities grows exponentially.

One of the most remarkable aspects of AI is its ability to continually improve through feedback. This concept, known as *reinforcement learning*, involves teaching algorithms to make sequences of decisions by rewarding desired behaviors and penalizing undesired ones. Think of it as a trial-and-error approach where the system learns the best actions to take to achieve a specific goal. This adaptability is especially valuable for developing technologies that cater to the unique needs of individuals with disabilities.

AI systems also rely heavily on vast amounts of data to function effectively. **Big Data** refers to the large volumes of structured and unstructured data generated by various sources, including social media, sensors, and healthcare records. The ability to analyze and draw insights from these datasets enables AI to make more accurate predictions and provide more personalized and effective solutions for users.

It's worth mentioning the role of ethics in AI development. The creation and implementation of AI systems must consider ethical issues such as privacy, bias, and fairness. AI must be developed and deployed in ways that respect individual rights and promote inclusivity. For example, ensuring that AI-driven assistive technologies are accessible to all individuals, regardless of socioeconomic status, is a key ethical concern.

The exploration of AI's concepts wouldn't be complete without acknowledging the importance of human-AI collaboration. AI is not intended to replace human abilities but to augment them. In the context of disabilities, this collaborative approach can lead to innovations that significantly enhance communication, mobility, and overall quality of life. By working together, humans and AI can achieve outcomes that neither could accomplish alone.

These basic concepts of AI lay the foundation for understanding how the technology can be harnessed to address the needs of individuals with disabilities. From machine learning and natural language processing to computer vision and robotics, each element contributes to creating a more inclusive and accessible world. As we delve deeper into specific applications and case studies in subsequent chapters, remember that the ultimate goal of AI is to empower human potential and improve lives, one innovation at a time.

This journey through the basics of AI is just the beginning. As we advance, we'll explore how these concepts translate into tangible benefits for people with different types of disabilities. Whether it's enhancing communication for those with speech impairments or providing navigation aids for the visually impaired, AI's potential to revolutionize accessibility is boundless. Stay tuned as we uncover the myriad ways AI is making a difference in the lives of millions worldwide.

History and Evolution of AI

Understanding the history and evolution of artificial intelligence is crucial to grasping its current and future potential, especially for enhancing the lives of people with disabilities. The story of AI doesn't just span a few innovative years; it stretches across decades and even centuries of human ingenuity and curiosity.

The genesis of AI can be traced back to ancient myths and stories that featured automated beings capable of simulating human intelligence. Early notions of artificial beings and mechanical devices appeared in mythology and literature, such as the ancient Greek tale of Talos, a giant automaton made by Hephaestus. This longing for artificial companions laid the groundwork for what would become a serious scientific endeavor centuries later.

Fast forward to the 20th century, where the concept of AI transitioned from myth to a legitimate scientific field. In 1950, Alan Turing, a British mathematician and logician, posed the question,

"Can machines think?" His seminal paper, "Computing Machinery and Intelligence," proposed the Turing Test, a method to determine if a machine could exhibit human-like intelligence. This marked a significant turning point, crystallizing the philosophical and technical aspects of AI.

The mid-20th century saw the formal birth of artificial intelligence as a field of study. In 1956, the Dartmouth Conference, organized by John McCarthy, Marvin Minsky, Nathaniel Rochester, and Claude Shannon, officially introduced the term "artificial intelligence." This event is widely considered the birth of AI as an academic discipline. The pioneers of this era envisioned machines that could perform tasks requiring human intelligence, such as learning, reasoning, language understanding, and problem-solving.

Early AI research focused on symbolic AI or "Good Old-Fashioned AI" (GOFAI), which used symbolic representations and rule-based systems to mimic human problem-solving. Programmers manually encoded knowledge and rules into these systems, which could solve algebra problems or prove mathematical theorems. However, these early AI systems struggled with scalability and adaptability. They were brittle, meaning they couldn't easily handle new or unexpected situations; if something fell outside their pre-programmed rules, they failed.

The 1980s introduced machine learning, shifting the focus from explicit programming to systems capable of learning from data. This period saw the rise of neural networks, inspired by the human brain's structure and function. Despite early enthusiasm, neural networks faced significant challenges due to limited computational power and training data, leading to a period known as the "AI Winter" where interest and funding in AI research waned.

The resurgence of AI came in the 2000s, driven by breakthroughs in computational power, the availability of large datasets, and advancements in algorithms. This era is often referred to as the "Deep Learning Revolution." Deep learning, a subset of machine learning, utilizes multilayered neural networks to automatically discover representations in data. This breakthrough enabled machines to achieve unprecedented levels of accuracy in tasks like image and speech recognition.

One of the landmark moments in AI history was in 2012 when a deep learning algorithm developed by Geoffrey Hinton and his team at the University of Toronto won the ImageNet Large Scale Visual Recognition Challenge by a significant margin. This demonstrated the potential of deep learning to tackle complex problems and led to a surge in interest and investment in AI research and applications.

While deep learning transformed many industries, AI's potential in assisting people with disabilities began to gain significant traction. Technologies like speech recognition, computer vision, and natural language processing opened new avenues for creating accessible solutions. For instance, screen readers for the visually impaired advanced significantly with the help of AI, providing more accurate and context-aware textual descriptions of images and web content.

Moreover, AI-driven applications have revolutionized communication aids for individuals with speech impairments. Predictive text and voice generation technologies, powered by machine learning models, allow for faster and more intuitive communication, breaking down barriers and fostering independence.

The evolution of AI hasn't been without its challenges and ethical considerations. Concerns around data privacy, algorithmic bias, and the digital divide persist. Ensuring that AI technologies are developed inclusively and ethically is crucial for maximizing their positive impact, particularly for marginalized communities, including people with disabilities.

As we navigate through the 21st century, AI continues to evolve at a breathtaking pace. Innovations like reinforcement learning, where AI learns to make decisions by interacting with its environment, and generative adversarial networks (GANs), which can create new data samples, are pushing the boundaries of what's possible. These advancements hold immense promise for developing even more sophisticated and personalized assistive technologies.

In summary, the history and evolution of AI is a story of relentless human creativity and innovation. From ancient myths to modern breakthroughs, AI has transformed from a philosophical question into a transformative technology. As we move forward, the continued evolution of AI promises to empower and elevate the lives of individuals with disabilities, creating a future where technology and human potential go hand in hand.

Chapter 3:
The Intersection of AI and Disability

As we venture into the intersection of AI and disability, it becomes clear that artificial intelligence holds transformative potential to level the playing field for individuals with disabilities. More than just a technological marvel, AI can eliminate barriers that once seemed insurmountable. Picture a world where voice-activated assistants translate spoken language into text instantly, empowering those with hearing impairments in real-time conversations. Imagine intelligent prosthetics that adapt fluidly to different terrains, giving freedom of movement back to those with mobility challenges. Yet, while the potential benefits are immense, it is crucial to be vigilant about risks, such as privacy issues and the equitable distribution of these technological advancements. Balancing optimism with caution ensures that AI can truly become a cornerstone in enhancing accessibility, independence, and quality of life for everyone.

The Role of AI in Accessibility

Artificial Intelligence (AI) is revolutionizing the landscape of accessibility for people with disabilities, embedding itself deeply into the fabric of daily life. At its core, AI aims to simulate human intelligence, and this capacity is being harnessed in increasingly inventive ways to bridge gaps in everyday experiences for those with disabilities. What once seemed like science fiction is now a reality, from voice-activated assistants reading out emails to advanced algorithms that can guide visually impaired individuals through complex environments.

Accessibility is more than just a convenient add-on; it is a fundamental right. The proliferation of AI technologies aims to ensure this right is afforded to all, irrespective of physical or cognitive limitations. One of the most prominent examples of AI in action is the broader spectrum of speech-to-text and text-to-speech applications that cater to individuals with hearing or speech impairments. When a meeting can be transcribed in real-time, barriers that once made professional participation difficult begin to dissolve.

For the visually impaired, AI offers navigation and independence unlike any other tool in history. Devices equipped with AI can interpret the environment, identify objects, read text aloud, and even detect obstacles. Imagine an AI system that can guide a person through a crowded subway system, narrating steps, and helping avoid hazards. These are not mere gadgets but lifelines that provide a significant leap toward independence.

Yet, it's not just about aiding independent travel. AI also personalizes the digital experience for people with disabilities. Customizable settings in smartphones and computers, driven by machine-learning algorithms, adapt to user preferences and needs, thus enhancing usability. Voice recognition systems like Apple's Siri, Google's Assistant, and Amazon's Alexa are tailored to recognize a broader range of speech patterns, including those affected by speech impairments.

However, the impact of AI in accessibility transcends beyond individual tools and applications. It fosters a more inclusive society by promoting equal access to information and services. Educational platforms, for instance, utilize AI to create adaptive learning environments. These systems analyze a student's interaction patterns and adapt content delivery to cater to individual learning needs. In doing so, they ensure that students with cognitive disabilities receive the same quality education, tailored to their capabilities and requirements.

Moreover, AI's role in accessibility is prominent in healthcare. Predictive analytics and diagnostic tools powered by AI can facilitate early detection of conditions that may lead to disabilities. For example, AI can analyze vast datasets to predict the risk of strokes or seizures, enabling timely intervention. Wearable devices that monitor vital signs and environmental factors also feed into AI algorithms, providing real-time feedback and alerts, thereby augmenting the autonomy of individuals with chronic health conditions.

In the workplace, AI-driven accessibility tools enable a level playing field. Automated transcription services, ergonomic workstation setups optimized through AI, and smart software that adapts to an employee's workflow can significantly enhance productivity and inclusivity. This not only benefits individuals with disabilities but also cultivates a more diverse and innovative workforce.

Yet, despite these advancements, the journey of integrating AI into accessibility is not without challenges. Ethical considerations, such as ensuring non-bias in AI algorithms and protecting user data privacy, are essential. AI systems must be trained on diverse datasets to avoid perpetuating existing biases, and stringent measures should be employed to secure personal data, especially for vulnerable populations.

Ensuring that AI applications are user-friendly and intuitively designed remains a high priority. Often, the best solutions arise when developers actively involve individuals with disabilities in the design process. This user-centered approach ensures that the technologies address real needs and preferences, resulting in more effective and widely accepted tools.

Research and development continue to push the boundaries of what's possible. Brain-computer interfaces (BCIs), for instance, allow users to control devices using their thoughts. This area, still in its nascent stages, holds promise for those with severe physical disabilities. Imagine a world where communication and control are facilitated by mere thought, thus breaking down the most formidable barriers of mobility and speech.

Furthermore, the potential of AI in enhancing accessibility is not limited to developed nations. Globally, initiatives are underway to ensure that AI-driven accessibility tools reach the farthest corners. By addressing language barriers and local challenges, AI can become a global equalizer, amplifying its impact manifold.

As AI continues to evolve, its role in accessibility will undoubtedly expand, opening new avenues for empowerment and independence. The ultimate goal remains clear: creating a world where disabilities don't define or limit an individual's reach or aspirations. This is the promise of AI—a promise that, if pursued with diligence and inclusivity, can truly transform lives.

Potential Benefits and Risks

The application of artificial intelligence (AI) in the context of disability holds transformative potential. By assessing its advantages and drawbacks, we can appreciate the multifaceted impact that AI technology has on accessibility, independence, and quality of life for individuals with disabilities. Let's dive into both the promising benefits and the significant risks that come with integrating AI into the daily lives of those with disabilities.

One of the most compelling advantages of AI is its ability to enhance accessibility. AI-powered tools, such as speech-to-text software and image recognition systems, make everyday tasks more manageable for people with disabilities. For instance, visually impaired individuals can navigate their environments more easily using AI-powered applications that describe surroundings through audio cues. Similarly, hearing-impaired individuals gain access to real-time transcription services that convert spoken words into text, breaking down communication barriers.

Independence is another major benefit. AI-driven assistive technologies empower individuals to perform tasks that they might not be able to accomplish on their own otherwise. From smart home devices that respond to voice commands to robotic limbs and exoskeletons that enhance mobility, AI presents ways for individuals to live more self-sufficiently. These tools provide significant autonomy, which immensely contributes to one's self-esteem and overall well-being.

Moreover, AI offers personalized support. Machine learning algorithms can adapt to the unique needs of each user, creating custom solutions that evolve over time. For example, educational platforms using AI can provide tailored learning experiences for individuals with cognitive disabilities, addressing specific challenges and improving cognitive functions through targeted exercises.

However, it is critical to acknowledge the risks that accompany these benefits. Privacy concerns are at the forefront. Many AI applications require extensive data collection to function optimally, raising questions about data security and user privacy. There is always a risk that sensitive information, such as medical history or personal habits, could be misused or leaked.

AI systems can also exacerbate existing inequalities. Not all individuals with disabilities have equal access to the latest technologies due to economic constraints. As such, the divide between those who can afford advanced assistive technologies and those who cannot may widen, creating new disparities in accessibility and independence.

Furthermore, the reliance on AI technologies introduces potential reliability issues. AI systems are not infallible; they may malfunction or provide incorrect outputs, which could be critical or even dangerous in some situations. For instance, a malfunctioning AI-powered wheelchair might result in accidents, underscoring the imperative need for rigorous testing and reliable support structures.

Inclusivity in AI development is another area of concern. If developers do not include individuals with disabilities in the design and testing phases, the resulting products may not fully address the targeted needs. This oversight can limit the effectiveness of AI solutions and may even result in technology that inadvertently creates new barriers.

Ethics play a crucial role in the discourse on AI and disability. The deployment of AI tools must be guided by principles that prioritize respect, consent, and the well-being of the end-users. Ethical dilemmas may arise regarding who gets access to these technologies and under what conditions they are used, which beckons an ongoing dialogue among stakeholders.

Finally, the rapid pace of AI development presents a challenge for regulatory frameworks. Keeping policies and regulations up to date with technological advancements is essential to ensure that AI tools are safe, equitable, and effective for all users. The potential for misuse of AI in surveillance or unauthorized monitoring also needs stringent oversight to protect the rights and freedoms of individuals with disabilities.

In conclusion, the intersection of AI and disability harbors immense promise but is also fraught with challenges. We must navigate this landscape with careful consideration, balancing the remarkable benefits against the inherent risks. By doing so, we can pave the way for a future where AI truly serves as a beacon of empowerment for

Inclusive Intelligence

individuals with disabilities, enhancing their lives in ways previously unimaginable.

Chapter 4:
AI-Powered Assistive Technologies

Imagine a world where technology weaves itself seamlessly into the lives of individuals with disabilities, enhancing their independence and quality of life. AI-powered assistive technologies are transforming this vision into reality. These innovations harness the power of artificial intelligence to break down barriers, offering personalized support and solutions that adapt in real-time. Whether it's through advanced speech recognition systems for those with communication challenges, intelligent prosthetics that move naturally, or smart home devices that cater to specific needs, AI is providing newfound autonomy and empowering individuals like never before. This chapter explores these groundbreaking technologies, underscoring their potential to revolutionize how we approach disability and accessibility. By integrating AI into assistive tech, we're not just creating tools; we're crafting lifelines that bring freedom, dignity, and opportunity to millions around the globe.

Overview of Assistive Tech

Assistive technology encompasses a broad array of tools and devices designed to enhance the lives of individuals with disabilities. These innovations aim to break down the barriers that hinder daily activities, communication, education, and employment. As technology advances, the landscape of assistive tech continuously evolves, offering new ways to support independence and improve quality of life. In this realm, AI-powered solutions are becoming increasingly invaluable, providing more intuitive, adaptable, and effective assistance than ever before.

In the past, assistive technologies primarily consisted of mechanical and electronic aids, such as wheelchairs, hearing aids, and text-to-speech devices. While these tools were revolutionary in their time, the integration of AI brings an entirely new dimension, allowing for more personalized and dynamic support. AI algorithms analyze vast amounts of data to understand user behavior and preferences, learning and adapting over time to deliver more effective and customized assistance.

One of the key strengths of AI in assistive technology is its ability to process and interpret complex data in real time. This real-time capability is crucial for applications such as navigation aids for the visually impaired, where immediate responses to changing environments can significantly enhance user safety and autonomy. Similarly, AI-powered hearing aids can filter background noise and focus on specific sounds, making conversations easier to follow in noisy environments.

Moreover, AI in assistive tech isn't just about enhancing existing tools—it's about creating entirely new possibilities. For example, brain-computer interfaces (BCIs) use AI to interpret neural signals, enabling individuals with severe physical disabilities to control devices and communicate using their thoughts. Such innovations represent a frontier in assistive tech, where the boundary between human capability and technological augmentation becomes increasingly blurred.

AI-driven assistive technologies also promote inclusivity by enabling better access to information and communication. Natural language processing (NLP) algorithms power speech recognition and generation tools that help individuals with speech impairments communicate more effectively. These tools can transcribe spoken words into text, suggest vocabulary, and even simulate natural conversation, bridging the gap between users and their environment.

The educational sector benefits substantially from AI-powered assistive tech as well. Personalized learning environments utilize machine learning algorithms to adapt to the unique needs of each student, providing customized lessons and resources. This approach helps students with cognitive disabilities to learn at their own pace, ensuring that they receive the support necessary to succeed academically. By making education more accessible, AI tools can open doors to opportunities that were previously out of reach for many individuals with disabilities.

AI-powered assistive technologies also extend their impact to the workplace. By automating routine tasks and providing intelligent support, these tools enable individuals with disabilities to perform their jobs more efficiently and independently. Adaptive software can adjust to different user abilities, ensuring that everyone has the tools they need to succeed. For instance, AI-driven screen readers help visually impaired employees navigate digital content, while voice-activated assistants facilitate communication and task management for those with mobility challenges.

Despite the myriad benefits, the implementation of AI in assistive tech is not without challenges. Ethical considerations, such as privacy and data security, are paramount when dealing with sensitive user information. Developers must ensure that AI systems are designed with fairness and inclusivity in mind, avoiding biases that could disadvantage certain groups of users. Additionally, the cost of cutting-edge technologies can be a barrier to widespread adoption, necessitating efforts to make AI solutions more affordable and accessible.

To maximize the potential of AI-powered assistive technologies, collaboration between diverse stakeholders is essential. This includes partnerships between tech companies, healthcare providers, educators, and advocacy groups, ensuring that the development and deployment of these tools are guided by a deep understanding of user needs. Government support and funding can also play a crucial role in advancing research and making these technologies available to those who need them most.

In summary, AI-powered assistive technologies represent a transformative force in the realm of disability support. By leveraging the capabilities of artificial intelligence, these tools offer unprecedented levels of customization, efficiency, and accessibility. As technology continues to evolve, the potential for AI to empower individuals with disabilities only grows, promising a future where barriers are continually diminished and opportunities for independence and participation are expanded.

Looking ahead, the ongoing integration of AI in assistive tech will likely lead to even more innovative solutions. Emerging technologies like augmented reality (AR) and virtual reality (VR) are already showing promise in developing immersive, interactive experiences tailored to users' needs. These advancements could revolutionize how individuals with disabilities engage with the world, both in physical and virtual spaces.

The future of AI-powered assistive tech is bright, filled with possibilities that can make a profound difference in the lives of millions. As we continue to explore and harness the power of AI, we must remain committed to creating inclusive, fair, and user-centered solutions. By doing so, we can ensure that everyone, regardless of their abilities, has the opportunity to thrive and participate fully in all aspects of society.

Case Studies and Examples

AI-powered assistive technologies have transformed the lives of individuals with disabilities across a multitude of dimensions. Imagine a world where ordinary tasks, previously laden with obstacles, become seamless thanks to the innovative application of artificial intelligence. This section highlights several compelling case studies and examples illustrating how AI is actively closing gaps in accessibility, boosting independence, and enriching quality of life for many.

Case Study 1: Voiceitt

Take the case of Voiceitt, an AI-driven speech recognition app designed for people with speech impairments. It's a game-changer, especially for those with conditions such as cerebral palsy, Down syndrome, or post-stroke speech impairments. Voiceitt's technology leverages machine learning algorithms to interpret non-standard speech patterns, translating them into comprehensible text or synthesized speech in real-time. This breakthrough not only facilitates clearer communication but also fosters greater social interaction and inclusion.

The feedback from users has been overwhelmingly positive. Many report feeling empowered and less isolated, seeing their self-esteem soar as they engage more confidently in daily conversations. For educators and healthcare professionals, tools like Voiceitt symbolize a leap forward in providing tailored support to speech-impaired individuals.

Case Study 2: Seeing AI

Next, consider Seeing AI, an application from Microsoft that helps visually impaired individuals navigate their surroundings. Utilizing advanced computer vision and natural language processing, Seeing AI can describe people, read text, identify products, and recognize currency—essential functions for daily living. Its user-friendly interface and voice-guided instructions make it accessible to a wide demographic.

Take Maria, for example, a visually impaired woman who expressed that Seeing AI has become her "virtual eyes." With AI's assistance, Maria confidently performs activities that once felt insurmountable, from reading her mail to shopping independently. The technology's profound impact highlights its potential for transforming the everyday experiences of visually impaired individuals worldwide.

Case Study 3: Liftware

In the realm of physical disabilities, Liftware's utensils offer an inspiring example of AI in action. These innovative eating tools, equipped with sensors and microprocessors, stabilize hand tremors to assist individuals with conditions like Parkinson's disease. The device's adaptive algorithms continuously adjust to the user's tremors, significantly improving their dining experience.

For someone like John, who has Parkinson's, Liftware's utensils provide a renewed sense of independence. No longer needing assistance with meals, John can dine with dignity, enhancing his overall quality of life and reducing caregiver burden. This case underscores the importance of personalized assistive technologies in fostering autonomy.

Case Study 4: ReWalk Robotics

ReWalk Robotics presents another powerful example, focusing on mobility solutions for individuals with spinal cord injuries. This exoskeleton, powered by AI, enables users to stand and walk with minimal assistance. It's more than just a medical device; it's a symbol of hope and resilience.

For veterans like Brian, who sustained a spinal cord injury during service, ReWalk has been miraculous. The ability to walk again and engage in upright activities has been life-altering, providing not just physical benefits but also emotional and psychological support. Stories like Brian's showcase the boundless potential of embedding AI in mobility aids.

Case Study 5: Cognoa's Autism Screener

In terms of cognitive disabilities, Cognoa's Autism Screener employs AI to facilitate early diagnosis and intervention. This tool analyzes behavior patterns from video clips and parent questionnaires, identifying signs of autism in children. Early and accurate diagnosis leads to timely interventions, crucial for improving long-term outcomes.

For families like the Thompsons, the screener has eased the arduous journey of seeking a diagnosis for their child. With Cognoa's insights, they accessed support services and began targeted interventions early, significantly enhancing their child's developmental trajectory. This example highlights AI's role in bridging diagnostic gaps and promoting proactive health management.

Case Study 6: Brain HQ

Brain HQ, on the other hand, provides cognitive training for individuals experiencing cognitive decline due to aging or brain injuries. This platform uses AI to tailor exercises to each user's unique cognitive profile, promoting improvements in memory, attention, and processing speed.

For seniors like Alice, who faced early signs of dementia, Brain HQ has been instrumental. With tailored exercises, Alice noticed marked improvements in her cognitive functions, allowing her to maintain her independence and enjoy a better quality of life. This case is a testament to how AI-powered cognitive aids can be a lifeline for many facing cognitive challenges.

To sum it up, these case studies reflect the diverse and transformative impact of AI-powered assistive technologies. From speech recognition and computer vision to adaptive utensils and cognitive training, the common thread is clear: AI is not just a tool but a game-changer in the landscape of accessibility. Each example embodies the promise of a more inclusive world where technology empowers everyone, regardless of their abilities.

The road ahead is paved with opportunities to innovate and refine these technologies further. As AI continually evolves, its applications in assistive technologies will undoubtedly expand, bringing even greater benefits to those who need them most. The journey to a fully accessible and inclusive society is ongoing, but with AI at the forefront, the future looks promising. These cases serve as a beacon of hope, illustrating what is possible when technology meets human potential in the most inclusive way.

Chapter 5:
Smart Assistants and Personal Aides

Smart assistants and personal aides are revolutionizing how individuals with disabilities navigate their daily lives, making technology work seamlessly to enhance independence and quality of life. These AI-powered tools, such as virtual assistants, can handle myriad tasks, from setting reminders to managing complex schedules, providing a level of personalized support previously unimaginable. Moreover, they facilitate improved communication and deeper interaction, breaking down barriers and opening new channels for engagement. By understanding and anticipating users' needs, smart assistants can offer tailored solutions, transforming challenges into manageable tasks. Integrating these tools into everyday routines exemplifies the transformative potential of AI, illustrating a future where technology isn't just a convenience but an empowering force that enables people to live their lives fully and autonomously. With continuous advancements, smart assistants will only become more intuitive and indispensable, driving us toward a more inclusive and accessible world for everyone.

Virtual Assistants for Daily Living

Virtual assistants, powered by sophisticated AI algorithms, are becoming indispensable tools for enhancing the lives of people with disabilities. These aren't just digital helpers; they're transformative companions that bring about a new level of independence and accessibility to daily tasks. Imagine a world where a spoken command can manage your calendar, control your home environment, or even assist in complex problem-solving. This is the realm of virtual assistants for daily living.

From setting reminders to orchestrating entire daily routines, virtual assistants like Amazon's Alexa, Google Assistant, and Apple's Siri have revolutionized how people interact with technology. For individuals with disabilities, these assistants offer more than convenience; they are lifelines to autonomy. A person with visual impairment can ask their assistant to read out emails, while someone with limited mobility can control smart home devices without lifting a finger. Such functionalities dramatically reduce the physical and mental effort required to perform everyday activities.

In addition to handling routine tasks, virtual assistants play crucial roles in enhancing communication. For example, those with speech impairments can benefit from AI that translates text into speech, allowing them to engage more fully in conversations. Similarly, people who are hard of hearing can leverage assistants to provide visual alerts for various auditory signals, thereby staying connected and informed. These applications not only break down barriers but also help in building a more inclusive society.

Moreover, virtual assistants are continually evolving, adapting through machine learning to better understand and respond to personalized needs. Utilizing user preferences and behavior patterns, these assistants can suggest actions or reminders before they are even asked, taking proactivity to new heights. Imagine waking up to find your assistant has already checked traffic conditions, ensuring you leave on time for an appointment. This kind of anticipatory support can significantly enhance the quality of life.

It's also essential to discuss the implications for mental health. Virtual assistants can offer prompts for medication, suggest relaxation exercises, and even check in emotionally, providing a sense of companionship. For individuals dealing with cognitive disabilities, these assistants can provide step-by-step guidance for tasks that might otherwise be overwhelming. Through consistent and empathetic interactions, verbal affirmations, and useful suggestions, virtual assistants can become pillars of emotional support.

Furthermore, the integration of virtual assistants with other AI technologies amplifies their effectiveness. When combined with wearable devices, for instance, they can monitor physiological indicators like heart rate and alert users and caregivers to any anomalies. This kind of real-time health monitoring is invaluable, providing not just convenience but also a safety net.

Accessibility extends to education as well. Virtual assistants can help students with disabilities manage their schedules, access educational content in accessible formats, or even assist with studying by posing quiz questions or summarizing notes. This level of support can bridge significant gaps in educational attainment, making learning more inclusive and accessible.

However, the true potential of virtual assistants lies in their customization. AI developers are increasingly focusing on building features that cater specifically to various disabilities. For instance, interfaces are being refined to better understand non-verbal cues for users who can't speak, or to provide more robust text-to-speech functionalities for those who find reading challenging. These tailored experiences ensure that the technology is not a one-size-fits-all solution but an adaptable tool tailored to individual needs.

Despite these advancements, there are challenges to consider. Privacy remains a significant concern, especially when dealing with sensitive health information. Ensuring data security and building trust with users is paramount. Developers and stakeholders must adhere to strict privacy guidelines to ensure that user data is protected and used ethically. Transparent data policies and robust encryption methods are essential components in this effort.

Cost also poses a barrier to wider adoption. High-quality virtual assistants and the smart devices they often control can be expensive. Efforts need to be made to subsidize these technologies or provide them through healthcare plans to ensure equitable access. Collaborations between tech companies, governments, and non-profits can play a crucial role in making these assistive technologies more affordable and widespread.

There's also the issue of technological literacy. Not everyone is comfortable using advanced technologies, and for some, this might be an additional barrier. Educational programs aimed at teaching users and caregivers how to effectively utilize virtual assistants can be invaluable. Simple, intuitive interfaces and comprehensive help guides can further ease this learning curve.

Moving forward, the potential for innovation is boundless. Developers are exploring the integration of natural language processing to create more conversational and intuitive interactions. With advancements in contextual understanding, these assistants will soon be able to carry out more complex tasks, making them even more indispensable.

Through collaborative efforts and continuous innovation, virtual assistants are primed to revolutionize daily living for those with disabilities. They embody the spirit of empowerment, facilitating independence, and enhancing quality of life. As technology progresses, these digital aides will become even more capable, personalized, and integral to the daily lives of countless individuals, proving that the future of accessibility is not just promising—it's here.

Enhancing Communication and Interaction

When it comes to leveraging the potential of AI for people with disabilities, the enhancement of communication and interaction stands out as one of the most transformative areas. It's not just about making life more convenient; it's about breaking down walls that have stood for far too long. Imagine a world where assistive technology allows individuals to engage in conversations without the barriers that previously held them back. This vision is no longer a distant dream but a reality rapidly unfolding before our eyes.

Smart assistants and personal aides have revolutionized the way we think about communication. Virtual assistants like Siri, Alexa, and Google Assistant are now equipped with features that do more than just answer questions or set reminders. For individuals with speech impairments, these smart assistants become invaluable tools. Innovations in speech recognition and natural language processing (NLP) allow these assistants to understand and interpret speech patterns that may not conform to typical speaking norms. This can be life-changing, as it opens up new avenues for interaction, both socially and professionally.

Consider a scenario where an individual with a speech impairment needs to make an urgent phone call but struggles with intelligible speech. Historically, making this kind of call could be daunting and possibly ineffective. However, with AI-driven personal assistants, they can now communicate clearly and effectively through text-to-speech conversion. The personal aide interprets their text input and translates it into a natural-sounding voice, ensuring the message is delivered accurately and respectfully. This is not just technology in action; it's empowerment.

In addition to speech recognition, AI has made significant strides in the field of facial recognition and emotional intelligence. For people who have difficulty interpreting social cues, such as those on the autism spectrum, AI can offer unparalleled support. Utilizing computer vision and emotion-sensing algorithms, smart assistants can provide real-time feedback on the emotional states of people in a conversation. They can alert the user if someone seems confused, amused, or upset, thereby offering cues that the user may otherwise miss. These tools pave the way for richer, more nuanced social interactions by providing a form of augmented social intuition.

Far from being limited to face-to-face interactions, these AI capabilities extend into digital communication as well. Email, instant messaging, and social media platforms can all benefit from AI enhancements. Take, for example, grammar and style correction tools powered by AI. They do more than just correct spelling errors; they suggest changes to improve clarity and tone. For people with learning disabilities or cognitive impairments, these tools can be game-changers. They mitigate the anxiety of written communication, making it easier and more accessible.

Interestingly, AI isn't just enhancing how we communicate; it's also broadening who we can communicate with. Language barriers are a significant challenge in our increasingly globalized society. Smart assistants equipped with real-time translation capabilities can bridge these gaps. Imagine attending an international conference where you don't speak the local language. With a smart assistant by your side, language ceases to be a barrier. These tools can translate speech in real-time, facilitating seamless, multi-lingual interaction.

Empathy-driven features in smart assistants are another area ripe with promise. AI has gotten better at understanding context and nuance, making it more capable of engaging in conversations that require a compassionate touch. This is particularly important in mental health support, where effective communication is crucial. AI-driven chatbots are being designed to offer preliminary mental health assessments and crisis intervention, providing a conversational partner during times of need. While these systems are not substitutes for human therapists, they can serve as a critical first line of support.

This brings us to the intersection of communication and interaction in the educational sphere. AI has a massive role to play here. Personalized learning platforms powered by AI can cater to the individual needs of students with disabilities, offering customized instructions and interactive lessons. For students who struggle with traditional forms of communication, these platforms can provide alternative means of expressing themselves, whether through visual aids, voice commands, or other adaptive technologies. The result is a more inclusive, effective educational experience.

Moreover, AI can enhance collaborative projects by offering tools that facilitate better teamwork among individuals with diverse abilities. For example, real-time transcription services can convert spoken words into text during a group discussion, making it easier for hearing-impaired participants to follow along. Collaboration platforms like Zoom and Microsoft Teams are already incorporating such features, making it clear that accessibility can go hand in hand with technological advancement.

One can't overlook the transformative potential of AI in public communication settings as well. Interactive kiosks equipped with AI can provide accessible information in real-time, whether it's directions in a busy airport, information about exhibits in a museum, or menus in a restaurant. These kiosks can use voice commands, touch screens with haptic feedback, and even sign language recognition to ensure they cater to the widest possible audience. Accessibility then becomes not just an afterthought but a fundamental design principle. In all these ways, AI is not just a tool but a bridge—connecting, enabling, and empowering individuals to communicate and interact in ways that were previously unimaginable.

AI-powered devices are also making strides in augmentative and alternative communication (AAC) systems, which are vital for individuals with speech and language disorders. These devices employ machine learning algorithms to predict and suggest words or phrases, thereby speeding up the communication process. By learning from the user's communication habits over time, these AI-driven AAC devices become increasingly efficient and personalized, making daily interactions smoother and more intuitive.

It's essential to highlight that the ultimate goal is not to replace human interaction but to augment it. Smart assistants and personal aides act as catalysts that facilitate more meaningful connections between people. They remove the friction and barriers that have previously made communication difficult or even impossible for many. The beauty of these AI technologies lies in their ability to adapt and learn, becoming more attuned to the needs and preferences of their users with each interaction.

Critically, the development of these smart assistants and personal aides needs to be inclusive from the ground up. User-centered design principles demand that people with disabilities be involved in the creation and testing of these technologies. Their input is invaluable to ensure that the end product meets real-world needs and addresses specific challenges effectively. Participation in the development process can also foster a greater sense of ownership and confidence in using these tools, further enhancing their positive impact.

As we look to the future, the evolution of AI promises even more sophisticated and nuanced tools for enhancing communication and interaction. We are on the cusp of seeing AI systems capable of understanding and responding to complex emotional and social dynamics. The potential for deep learning and advances in natural language processing means that these smart assistants will become even better at contextual understanding, making interactions feel more natural and intuitive.

In conclusion, enhancing communication and interaction through AI technologies is not merely a technical challenge; it's a profoundly human endeavor. It's about ensuring that every individual, regardless of their abilities, has the opportunity to share their voice, their thoughts, and their stories. By breaking down barriers and building bridges, AI can foster a more inclusive world where everyone can participate fully and meaningfully. As we continue to innovate and evolve, the promise of smarter, more empathetic, and more inclusive communication tools lies within

Chapter 6:
Vision and Hearing Impairment Solutions

In this chapter, we explore the transformative impact of artificial intelligence on addressing vision and hearing impairments, presenting solutions that were once considered the realm of science fiction. AI-driven technologies are now offering unprecedented opportunities for individuals with these impairments to navigate the world with greater ease and confidence. From smart glasses that translate visual data into auditory cues for the visually impaired, to advanced hearing aids that use machine learning algorithms to filter out background noise, AI is unlocking new levels of accessibility and independence. These innovations are not merely incremental improvements; they're fundamental shifts that enhance quality of life and empower individuals to engage more fully in their communities. Through practical applications and forward-thinking solutions, AI is proving to be a powerful ally in breaking down barriers and opening up a world of possibilities.

AI for the Visually Impaired

Vision impairment can pose significant challenges in day-to-day life, affecting one's ability to navigate environments, perform tasks, and engage with written and visual information. Artificial Intelligence (AI) has stepped in as a transformative ally, offering innovative solutions to enhance the lives of those with visual impairments. The integration of

AI-driven tools and devices opens up new avenues for greater accessibility, independence, and empowerment.

One prominent category of AI solutions lies in computer vision technology. Computer vision allows devices to "see" and interpret visual data, significantly aiding individuals with low vision or blindness. For example, AI-powered smartphone apps can provide real-time descriptions of a user's surroundings, detect objects, read texts aloud, and even recognize faces. By leveraging the capabilities of machine learning algorithms, these apps continuously improve their accuracy and usefulness, enabling users to more confidently interact with their environment.

Moreover, AI has revolutionized the realm of wearable technology. Smart glasses equipped with AI can identify objects, read text, and convey crucial visual information through audio outputs. These devices can connect to cloud-based AI systems to offer continuous updates and enhancements, ensuring users benefit from the latest advancements in technology. This seamless integration of AI-enhanced wearables fosters a greater sense of freedom and security, making the world more navigable and comprehensible for visually impaired individuals.

Another groundbreaking application of AI is in the realm of navigation. GPS-based mobile applications enhanced with AI features can provide detailed, voice-guided directions tailored specifically for visually impaired users. These apps not only guide users from point A to point B but also offer information about obstacles, traffic conditions, and nearby points of interest. Integrating data from various sensors, including cameras and accelerometers, makes these solutions more intuitive and responsive to real-world conditions.

Finally, there is a growing focus on AI for educational and professional environments. AI-driven text-to-speech engines and screen readers offer visually impaired individuals the ability to access digital content effortlessly. These tools process written content, including web pages, documents, and emails, converting them into natural-sounding speech. By incorporating advanced natural language processing, these systems offer more contextually accurate readings, enhancing comprehension and engagement.

Accessibility in education is further supported by AI-facilitated transcription services for visual content such as charts, graphs, and diagrams. By converting visual information into richly detailed audio descriptions, students with visual impairments can gain an equivalent understanding of complex subjects. As this technology evolves, its applications in diverse educational settings expand, fostering a more inclusive learning environment for all.

AI's role extends to improving accessibility in public spaces. For instance, AI-enabled kiosks and terminals in public transportation, airports, and shopping centers can provide audio guidance and descriptions of the services available. This significantly eases the stress of navigating unfamiliar spaces and ensures that visually impaired individuals have equitable access to public amenities.

Social inclusion and interaction also benefit from AI innovations. Tools like AI-based social robots can engage with visually impaired users, providing companionship and supporting daily activities. These robots use facial recognition and voice interaction capabilities to assist users in identifying and responding to social cues, thus fostering a more connected and inclusive social experience.

Despite these advancements, the road to perfecting AI for the visually impaired is ongoing. Developers face challenges like ensuring privacy, avoiding biases within AI systems, and addressing the diverse needs of users. Nevertheless, collaboration between technologists, healthcare professionals, and the visually impaired community is driving progress. User feedback is instrumental in refining these technologies, ensuring they meet the practical needs and preferences of their intended users.

AI's potential to improve the quality of life for visually impaired individuals is immense. By harnessing the power of artificial intelligence, we move closer to a world where visual impairments pose fewer barriers to independence and achievement. Innovations in this field reflect a broader commitment to inclusivity and accessibility, heralding a future where technology empowers everyone, regardless of their physical abilities.

AI for the Hearing Impaired

Artificial intelligence (AI) is revolutionizing the way we approach hearing impairments, creating powerful solutions that offer tremendous empowerment for individuals. Historically, the hearing-impaired community has faced significant barriers in communication, education, and social interactions. However, AI is now leveling the playing field. From enhancing traditional hearing aids to developing groundbreaking communication tools, AI is transforming challenges into opportunities.

A critical advancement is the evolution of smart hearing aids. Unlike their traditional counterparts, these devices use AI algorithms to adapt to various listening environments automatically. Imagine a hearing aid that not only amplifies sound but also distinguishes between background noise and conversations. This capability enables users to follow conversations in noisy places like restaurants, a task that was previously daunting. Furthermore, AI-driven hearing aids can be synced with smartphones, providing greater control through intuitive apps.

Another remarkable innovation is real-time translation and transcription services. AI-powered applications like Google's Live Transcribe convert spoken words into text in real-time, allowing those with hearing impairments to participate in conversations seamlessly. These apps can be a game-changer in educational settings, workplaces, and social gatherings, ensuring that no one is left out of important discussions.

Sign language interpretation is another area where AI shines. Traditionally, live interpreters or prerecorded videos have been the primary resources for translating spoken language into sign language. However, AI is now capable of offering real-time sign language interpretation through software solutions. Leveraging machine learning, these systems analyze a user's signs and convert them into text or speech instantly, making communication more fluid and reducing the need for human interpreters in everyday interactions.

AI isn't just about improving existing solutions; it's also about creating entirely new ways to communicate. Take the example of brain-computer interfaces (BCIs) which can interpret neural signals directly and convert them into speech or text. While still in its early stages, this technology holds promising potential for individuals who struggle with both hearing and speaking, providing them with an innovative communication channel.

Educational technology has also greatly benefited from AI. Tools like AI tutors and personalized learning systems can tailor educational content to suit the individual needs of hearing-impaired students. These systems use a combination of visual cues, text, and adaptive learning techniques to deliver a more inclusive educational experience. By adjusting to the pace and learning style of each student, AI can bridge educational gaps, ensuring that hearing-impaired students receive the same quality of education as their peers.

Video content, which is central to modern communication and education, becomes more accessible through AI as well. Captioning services have significantly improved, thanks to advanced speech recognition algorithms. These AI-driven captioning systems can swiftly generate accurate subtitles for live broadcasts, online courses, and even social media content. This development ensures that the hearing-impaired community has equal access to multimedia information and entertainment.

Moreover, wearable technologies are integrating AI to provide discreet yet powerful solutions. Smartwatches, for example, can be equipped with AI-powered speech-to-text applications, allowing users to receive transcribed conversations right on their wrists. These multipurpose devices can also provide alerts through vibrations or visual notifications, ensuring that users don't miss important auditory cues like alarms or phone calls.

But the impact of AI extends beyond practical applications; it fosters social inclusion. Consider social robots equipped with AI that can engage in lip reading, sign language, and even facial expression recognition. These robots can serve as companions and aides, making social interaction more accessible and enjoyable for individuals with hearing impairments. By bridging the communication gap, these robots can reduce feelings of isolation and loneliness.

Nonetheless, it's essential to address the ethical considerations involved. Ensuring that AI solutions are designed with the input of the hearing-impaired community is crucial. User-centered design approaches should prioritize accessibility, privacy, and usability. Developers must be mindful of the potential biases in AI algorithms that might inadvertently disadvantage users. Transparency in how AI systems work and the continuous involvement of the hearing-impaired community in development processes can mitigate these risks.

Furthermore, affordability and accessibility of AI technologies remain significant challenges. While AI-powered solutions offer much promise, their high costs can be prohibitive. Many individuals and families may find these technologies unattainable without adequate funding support from governments, nonprofits, and private sectors.

Hence, there is a pressing need for inclusive policies that subsidize AI solutions, ensuring equitable access for all.

It's also worth noting how AI can support mental health for those with hearing impairments. Communication barriers often lead to social isolation, which can affect mental well-being. AI-driven platforms offering virtual support groups or mental health resources tailored for the hearing-impaired can provide invaluable assistance. These platforms can facilitate community-building and foster a sense of belonging, mitigating the loneliness that many face.

Looking ahead, the possibilities for AI in assisting the hearing-impaired are vast. Continued advancements in machine learning, natural language processing, and wearable technology will undoubtedly contribute to more sophisticated and affordable solutions. As researchers and developers push the boundaries of what AI can achieve, we can expect increasingly personalized and intuitive tools that enhance the quality of life for those with hearing impairments.

In summary, AI's role in addressing hearing impairments is transformative. Through smart hearing aids, real-time transcription services, AI-driven sign language interpretation, brain-computer interfaces, and more, technology is dismantling communication barriers. By ensuring these innovations are both accessible and inclusive, we move towards a future where hearing impairments pose fewer obstacles to living a fulfilling, connected life. The confluence of AI and human ingenuity heralds a new era of empowerment and inclusivity.

Chapter 7:
Mobility and Physical Disabilities

Imagine a world where barriers to mobility dissolve with the wave of innovation in artificial intelligence. AI technologies are not just enhancing mobility solutions but revolutionizing them, making previously unimaginable levels of independence achievable for individuals with physical disabilities. From advanced robotic prosthetics that adapt to a user's unique movement patterns, to AI-driven exoskeletons that provide the strength and stability needed to walk, the possibilities are transformative. These innovations enable greater freedom in everyday activities, empowering people to live with a newfound confidence and autonomy. By focusing on personalized, adaptive solutions, we not only address the practical challenges but also reinforce the dignity and self-reliance of those navigating the world with physical disabilities. The promise of AI is not just in its technological prowess but in its potential to create a more inclusive society where mobility is accessible to all.

AI in Mobility Solutions

The marriage of AI and mobility solutions is nothing short of remarkable, especially when you consider its transformative impact on the lives of individuals with physical disabilities. The advent of AI has paved the way for a new era of accessibility, where barriers to mobility are not just reduced but, in many cases, eradicated. From intelligent prosthetics to self-navigating wheelchairs, AI is pushing the boundaries of what technology can accomplish.

One of the most significant advancements in this field is the creation of AI-powered prosthetics. Traditional prosthetics have seen limited functionality and required manual adjustments. With AI, however, these devices can now offer personalized, adaptive, and intuitive control. For example, sensors and machine learning algorithms can analyze the user's muscle signals and movements, enabling real-time adjustments for a more natural and seamless experience. It's this fine-tuning that transforms a once rigid device into something that feels like an extension of the body.

But the innovations don't stop at prosthetics. AI is also revolutionizing wheelchairs. Smart wheelchairs that integrate AI technologies can navigate through complex environments autonomously. These wheelchairs often use computer vision and LIDAR to interpret their surroundings, identifying obstacles and plotting the safest, most efficient route. This capability isn't just a novel addition; it's a life-changer for those who might find it difficult to navigate traditional wheelchair controls.

Moreover, AI is stepping into the realm of exoskeletons. These wearable robotic suits assist users with weakened or non-functional limbs to regain mobility. Equipped with sophisticated AI algorithms, these exoskeletons can predict the user's movements and provide the necessary mechanical assistance. This not only helps in walking but also in performing other essential daily activities, drastically improving the quality of life.

Yet, it's not just about physical devices. AI in mobility solutions also shines through in the software realm. For individuals with physical disabilities, planning and navigation through urban environments can pose numerous challenges. Here, AI-driven applications can provide invaluable support. For instance, apps that use real-time data to suggest the most accessible routes or to offer alerts about temporary obstructions or closures provide a new level of autonomy for users.

Transportation is another critical area where AI is making incredible strides. Autonomous vehicles are on the horizon, and their potential to transform mobility for people with disabilities is immense. These vehicles can offer door-to-door service without the need for a human driver, thereby eliminating one of the most significant barriers to independent travel. Though still in developmental and testing phases, the promise of autonomous vehicles is a beacon of hope for future mobility solutions.

Voice technology is an often-overlooked yet vital part of AI in mobility solutions. Voice-activated systems can allow users to control their environment effortlessly. Whether it's adjusting the settings on a smart wheelchair or instructing an autonomous car, voice technology makes these tasks simpler and more intuitive. This is critical for people whose mobility impairments also affect their ability to use traditional control mechanisms.

In parallel, immersive technologies like Virtual Reality (VR) and Augmented Reality (AR) are also leveraging AI to offer new dimensions in mobility solutions. VR can be used in rehabilitation settings to simulate walking and movement scenarios, providing valuable training to individuals in a safe and controlled environment. AR, on the other hand, can assist in navigation by overlaying critical information onto the user's real-world view, such as directions or alerts about obstacles.

It's essential to recognize that these AI innovations go beyond mere convenience. They significantly enhance the independence of people with physical disabilities, empowering them to live fuller, more autonomous lives. The ability to move freely and confidently impacts everything from employment opportunities to social interactions, making mobility not just a physical necessity but a cornerstone of overall well-being.

However, the journey of integrating AI into mobility solutions is not without challenges. One primary concern is the affordability and accessibility of these advanced technologies. While some flagship products demonstrate what's possible, they remain out of reach for many due to high costs. Thus, a critical focus for future development is making these technologies more accessible to a broader audience.

Another concern revolves around the reliability and security of AI systems. Given the crucial role these technologies play in individuals' lives, ensuring that they are both reliable and secure is paramount. Rigorous testing, quality assurance, and robust cybersecurity measures are essential components in this space. Failures or breaches could have serious, even life-threatening, consequences.

Ethical considerations also come into play. AI systems must be designed and implemented to respect users' dignity and rights. This involves inclusive design principles, ensuring that all aspects of the technology— from development and testing to deployment and maintenance— involve the target users. Their feedback is invaluable in creating solutions that are not only functional but also truly user-centered.

For mobility solutions to continue to evolve, collaboration between various stakeholders is crucial. This includes tech companies, healthcare providers, researchers, policymakers, and, importantly, the users themselves. By working together, they can identify gaps, brainstorm innovative solutions, and create technologies that are practical, effective, and inclusive.

The future of AI in mobility solutions is incredibly promising. As technologies continue to advance, we can expect even more sophisticated, intuitive, and accessible solutions to emerge. With each innovation, we move closer to a world where physical disabilities are no longer seen as insurmountable barriers but as challenges that can be met with ingenuity and compassion.

In summary, AI is revolutionizing the way we approach mobility for individuals with physical disabilities. Through intelligent prosthetics, autonomous wheelchairs, exoskeletons, and advanced software applications, the boundaries of what's possible are being continually pushed. These technologies not only enhance physical capabilities but also provide a profound sense of independence and dignity. As we forge ahead in this exciting frontier, the collective effort to make these advancements widely accessible and ethically sound will define the true success of AI in mobility solutions.

Robotic Assistance and Adaptability

In the realm of mobility and physical disabilities, the advent of robotic assistance is nothing short of transformative. From exoskeletons that help paralyzed individuals walk again to robots that aid in routine tasks, the spectrum of robotic technologies is vast and continually expanding. Central to this evolution is adaptability—robots must be capable of adjusting to the unique needs of each user, ensuring the highest level of functionality and independence.

Let's start with exoskeletons. These wearable robotic devices, initially developed for military and industrial applications, have found a profound purpose in healthcare. For individuals with spinal cord injuries or neurological disorders, exoskeletons can restore the ability to stand and walk. Beyond the physical benefits of re-engaging these muscles, the psychological impact is tremendous. Regaining mobility—even partially—enhances quality of life and fosters a sense of normalcy.

Adaptability in robotic assistance is not just about physical adjustments; it extends deeply into the software realm. Intelligent algorithms enable these robots to learn from user behavior, customize their responses, and improve over time. This constant learning and adaptation cater to the dynamic needs of evolving conditions, whether temporary, such as recovering from an injury, or permanent, like progressive muscular dystrophy.

The importance of user-centered design cannot be overstated. Robots like the JACO robotic arm, mounted on a wheelchair, exemplify how devices can be tailored to user requirements. Such robots can assist with everyday tasks—reaching for objects, opening doors, or even feeding—facilitating independence. Customization settings that allow the arm's speed, strength, and range of motion to be fine-tuned ensure that the device is neither too aggressive nor too slow, striking a balance that maximizes usability and comfort.

For children with physical disabilities, robotic assistance is making inroads in educational settings. Integrating robots in classrooms allows for inclusive participation in activities. For instance, a child using a robotic hand can engage in art projects or experiments alongside peers, fostering a sense of belonging and boosting self-esteem. Robotic playmates that can adapt to a child's physical abilities also show promise, offering companionship while serving as therapeutic tools.

Considering the aging population, robots are becoming indispensable. Aging individuals often face multiple challenges that compromise their independence. Robotics that assist with mobility, such as autonomous wheelchairs, are equipped with sensors and AI to navigate safely through homes and public spaces. These systems adapt to user preferences, learning common routes and navigating obstacles. For the elderly with degenerative conditions, adaptive robots that can modify their assistance based on the day-to-day variations in the user's physical capabilities are critical.

In addition to mobility, robotic assistance extends to home automation. Think of smart homes where robots assist in various chores tailored to the user's physical limitations. These robots can be connected to other smart devices in the home, allowing for seamless integration. For example, a robot vacuum that knows which rooms are frequented by the user and cleans those areas more meticulously showcases the harmony between tasks and user needs.

The robotics field is also tackling the issue of robustness and reliability. When it comes to critical assistance devices, downtime isn't an option. Advances in battery technology, durable materials, and fail-safe systems ensure that these robotic aids are trustworthy and resilient. Regular updates and preventative maintenance, powered by AI, predict and address potential failures before they become a problem.

Addressing the ergonomic needs through adaptable interfaces is another critical area of focus. Innovations like adaptive joysticks, voice control, and even EEG-based controls are expanding how users interact with their robotic assistants. These methods ensure that even those with significant motor control issues can effectively command their robotic aids, bridging a previously insurmountable gap.

Communities and users are increasingly involved in the co-development of these robots. Their firsthand experience and feedback are invaluable, leading to designs that are not only functional but also user-friendly. By building robots that are co-created with their intended users, engineers and designers ensure that they truly meet the diverse needs and preferences of the people they are meant to help.

Robotic assistance isn't limited to physical support. Emotional and social aspects are integral to holistic well-being, and robots are making strides in these areas as well. Social robots, equipped with AI, can detect signs of distress, offer companionship, and even alert caretakers in emergencies. These robots adapt to emotional cues and interactions, creating a supportive environment that goes beyond mere functionality.

Telepresence robots are another innovation improving the lives of those with severe physical disabilities. These devices enable individuals to have a virtual presence in places they physically can't go, be it a classroom, office, or family gathering. Equipped with cameras, microphones, and screens, they provide a means to interact just as a person physically there would.

The future of robotic assistance is filled with potential. Emerging technologies like soft robotics, which use flexible materials that mimic muscle movements, add a new level of adaptability and safety. These soft robots are better suited for interacting with humans and can provide more natural support and movements, reducing the risk of injury.

Another promising area is brain-computer interfaces (BCIs). These interfaces can interpret neural signals and convert them into commands for robotic devices, offering a direct control mechanism for users with severe physical impairments. While still in its infancy, BCI holds the potential to revolutionize how we think about robotic assistance, offering near-instantaneous adaptation to a user's mental commands.

Despite the technological advances, affordability and accessibility remain significant barriers. Efforts are underway to make robotic assistance devices more affordable through mass production and open-source designs. Funding from government and non-profits also plays a pivotal role in making these life-changing devices available to those in need.

It's inspiring to see robotic assistance breaking down barriers and opening new possibilities for people with physical disabilities. The adaptability of these devices ensures they are not just temporary solutions but companions for life, growing and evolving alongside the people they assist. By continuing to focus on user-centered design and leveraging the latest in AI and robotics, we can create a more inclusive

Maya Jordan

world where mobility challenges are not obstacles but opportunities for innovation and empowerment.

Chapter 8:
Cognitive Disabilities and AI

Artificial Intelligence is revolutionizing the lives of individuals with cognitive disabilities by offering tools that dramatically enhance cognitive support, learning, and memory. These AI-powered solutions not only provide personalized assistance but also adapt to the unique needs of each user, fostering a more inclusive environment. From apps that aid in task management and scheduling to intelligent tutoring systems that adapt to a student's learning pace, AI technologies empower individuals by reducing everyday challenges. The practical application of AI in cognitive support is not just a technological advancement; it's a step towards independence and improved quality of life for millions. By breaking down barriers, AI opens doors to unprecedented opportunities for cognitive development and social inclusion, making it possible for everyone to achieve their full potential regardless of cognitive limitations.

AI Tools for Cognitive Support

As we explore the transformative potential of AI in the realm of cognitive disabilities, it's crucial to begin by understanding the specific challenges individuals with cognitive disabilities encounter. Cognitive disabilities encompass a broad spectrum of conditions, including autism, Down syndrome, traumatic brain injury, dementia, and learning disabilities. These conditions can significantly impact memory, attention, problem-solving, language skills, and other cognitive functions. Artificial Intelligence (AI) has emerged as a formidable ally, offering tools that support cognitive function and improve quality of life. Let's delve into some of these tools and understand how they're making a difference.

One of the most common applications of AI in cognitive support is through *adaptive learning platforms*. These platforms use machine learning algorithms to create personalized learning experiences tailored to the unique needs of each individual. By analyzing a user's learning style and progress, these systems can adjust the difficulty level of tasks, provide instant feedback, and suggest personalized learning pathways. Platforms like Smart Sparrow and DreamBox are notable examples that employ AI to make education more accessible and effective for those with cognitive disabilities.

Another significant development in this space is AI-powered *memory aids*. Individuals with cognitive impairments often struggle with memory-related challenges, making it difficult for them to manage daily activities independently. AI tools like smart calendars, reminder apps, and voice-activated assistants can be lifesavers. Apps such as Google Calendar and Amazon Alexa can assist in remembering appointments, tasks, and even medication schedules. They offer prompts and reminders that help individuals maintain routines and manage their time more effectively, promoting greater independence.

Beyond memory aids, AI can also play a crucial role in enhancing *communication skills*. For those with conditions such as autism or speech disorders, engaging in social interactions can be daunting. AI-driven communication tools like augmentative and alternative communication (AAC) devices are designed to assist these individuals. These devices leverage natural language processing and machine learning to help users formulate and convey their thoughts more clearly. Proloquo2Go and Predictable are popular AAC apps that facilitate communication for individuals with speech and language difficulties, enabling them to express themselves and interact with others more confidently.

Virtual Reality (VR) and Augmented Reality (AR), augmented by AI, are also stepping up as potent tools for cognitive rehabilitation and skill training. VR and AR applications can create immersive environments for therapeutic and educational purposes. For instance, VR packages like RehAtt™ can create simulated environments to help individuals practice real-world activities and improve their cognitive function. These tools are especially beneficial for those recovering from brain injuries or living with neurodegenerative conditions, as they provide a safe space for cognitive exercises and skill development.

AI is also making strides in *behavioral monitoring and intervention*. Behavioral challenges are not uncommon among those with cognitive disabilities. Wearable devices and monitoring systems that utilize AI can track an individual's behavior patterns and predict potential issues before they escalate. For example, AI can analyze data from sensors and wearable devices to detect early signs of agitation or stress and alert caregivers. DrOmnibus's ABA DrOmnibus is an AI-based tool designed for Applied Behavior Analysis (ABA) therapy, offering data-driven insights and customized intervention plans to manage and improve behavioral outcomes.

Another fascinating application is AI's ability to personalize *social stories* and scripts, especially useful for individuals on the autism spectrum. Social stories are short descriptions of particular situations, events, or activities, which include specific information about what to expect and why. AI can help create personalized social stories that are tailored to an individual's experiences and needs, making it easier for them to navigate social scenarios. Through natural language generation, these AI tools can produce customized, contextually relevant stories that assist users in understanding and engaging in social interactions more effectively.

In the realm of early intervention, AI tools are proving invaluable. Early detection and intervention can markedly improve the outcomes for children with cognitive disabilities. AI-driven diagnostic tools and platforms can analyze a variety of data points to identify signs of cognitive impairments at an earlier stage than traditional methods. An example of this is the use of AI in analyzing speech patterns among children to detect early signs of autism. By providing early diagnosis, these tools pave the way for timely and more effective intervention strategies, enhancing developmental prospects.

Alignment with the Internet of Things (IoT) is yet another frontier where AI tools for cognitive support are creating ripples of change. The integration of AI with IoT devices can create a smart environment that responds to the needs of individuals with cognitive disabilities. Smart home systems equipped with AI can help automate daily tasks, provide environmental controls, and ensure safety. For instance, smart sensors can detect if a stove has been left on or if a door is unlocked and then prompt the individual or a caregiver accordingly. This interconnected environment fosters greater autonomy and safety for individuals with cognitive disabilities.

In addition to addressing practical and daily living challenges, AI tools can also support *cognitive therapy and mental health*. Tools like Woebot and Replika, which are AI-driven therapeutic chatbots, provide emotional support and cognitive-behavioral therapy (CBT). These tools can be particularly beneficial for individuals with cognitive impairments who may also experience anxiety, depression, or other mental health challenges. By offering a non-judgmental and always-available platform for interaction, these AI companions can help individuals better manage their emotional wellbeing.

Moreover, AI tools for cognitive support are not limited to individual use; they also provide valuable assistance to caregivers and educators. AI-driven analytics and reporting can offer insights into the progress and challenges faced by individuals with cognitive disabilities, helping caregivers and educators tailor their support strategies. Platforms like BrainHQ provide cognitive training programs but also offer progress tracking features that can be indispensable for those supporting individuals with cognitive disabilities, enabling a more informed and adaptive approach to care and education.

It's equally important to highlight the role of community and collaboration in the development and implementation of these AI tools. The involvement of users, caregivers, educators, and healthcare professionals in the design and deployment of AI technologies ensures that these tools are more attuned to the actual needs and contexts of those they are meant to serve. Community-driven platforms and open-source AI projects are fostering innovation that is not only inclusive but also practical and effective. When the voices of those with cognitive disabilities are heard, AI tools can be far more impactful and meaningful.

While the potential of AI tools for cognitive support is vast and growing, it's essential to navigate this landscape with a keen awareness of ethical considerations. Issues such as data privacy, transparency, and the potential for bias must be addressed proactively. Ensuring that AI tools are designed and deployed ethically is not just about compliance but about building trust and ultimately improving the lives of individuals with cognitive disabilities. When designed thoughtfully and deployed responsibly, AI can be a powerful catalyst for change, enhancing accessibility and transforming lives.

As we move forward, the integration of AI in cognitive support will likely become even more sophisticated and nuanced. Emerging technologies such as brain-computer interfaces (BCIs) and advanced neuroimaging, coupled with AI, hold promise for groundbreaking advancements in cognitive support. These innovations can

Improving Learning and Memory

Artificial intelligence (AI) offers transformative opportunities for enhancing learning and memory, especially for individuals with cognitive disabilities. By leveraging advanced algorithms and machine learning, AI systems can cater to unique cognitive needs, often providing personalized learning experiences that traditional methods fail to deliver. Visualize a world where the intricacies of one's learning process are seamlessly understood and addressed by smart systems designed to foster cognitive growth.

One of the most promising aspects of AI in this area is its ability to adapt to the individual's learning pace. Traditional educational systems often struggle to accommodate diverse learning speeds and styles. However, AI-driven educational tools can analyze how a person interacts with instructional materials, adjusting the content complexity and presentation format in real-time. This not only ensures that individuals do not lag behind but also helps in building confidence, reducing frustration, and sustaining engagement.

Memory enhancement—both in terms of retention and recall—is another critical area where AI can make a significant impact. For individuals with cognitive disabilities, remembering daily tasks or academic information can be incredibly challenging. AI applications, such as cognitive training apps, deploy games and exercises designed to stimulate neural pathways associated with memory. Over time, these tools can demonstrate tangible improvements in focus, problem-solving skills, and memory recall.

Speech recognition and natural language processing (NLP) technologies serve invaluable roles in improving learning and memory. By converting spoken language into text, AI tools help individuals who face difficulties in writing or typing adapt to learning environments without feeling hampered by their impairments. Consequently, learners can review material at their convenience, aiding in better information retention.

While AI's capabilities are impressive, the implementation of effective AI systems necessitates a thorough understanding of the learners' needs, strengths, and areas requiring development. Comprehensive assessments should precede AI integration, ensuring that the chosen tools are optimized for individual requirements. For example, children with dyslexia might benefit more from text-to-speech functions combined with visual aids, while adults with early-stage dementia may find memory games and cognitive reminders more beneficial.

Interactive AI platforms offer opportunities for repetitive, incremental learning—a method proven effective for cognitive reinforcement. These platforms can present information in various formats—text, audio, video—enabling repeated exposure in ways that are engaging rather than monotonous. Repetitive, incremental learning allows individuals to build upon what they already know, gradually integrating more complex concepts at a pace tailored to their specific learning thresholds.

In the classroom, AI can offer support to educators by analyzing student performance data and generating insightful analytics. This data-driven approach enables teachers to identify areas where a student might struggle, thereby allowing for planned interventions and targeted instructions. Importantly, the educator's role is complemented rather than replaced by AI, ensuring that the human touch remains integral to the learning experience.

Another significant advancement is the development of virtual reality (VR) environments tailored to support cognitive development. VR can immerse learners in simulated scenarios where they can practice task management, social interaction, and problem-solving in a controlled yet dynamic environment. These experiential learning opportunities often lead to better strategy formulation, cognitive agility, and practical knowledge application.

Customized AI-based learning applications can include a suite of tools designed to manage cognitive load effectively. For example, by breaking down large chunks of information into smaller, more digestible units, these applications help in better assimilation and retention of information. Furthermore, periodic assessments built into these programs can help in tracking progress and readjusting learning strategies as needed.

The importance of accessibility in these AI tools cannot be overstated. Ensuring that software platforms are user-friendly and inclusive of diverse needs is a prerequisite for effective learning. Features such as voice commands, flexible interface adjustments, and intuitive design are essential to cater to a broad spectrum of cognitive impairments. These functionalities should be ingrained within the system architecture, allowing for seamless user interactions.

AI-driven systems also enable the creation of personalized memory aids. For individuals who struggle with remembering appointments, deadlines, or essential tasks, AI can provide cognitive prosthetics such as digital reminders, smart calendars, and automated to-do lists. These tools not only assist in task management but also help minimize the psychological stress associated with forgetfulness.

AI's role in facilitating cognitive resilience deserves attention. By providing continuous stimulation and adaptable learning environments, AI systems can contribute to neuroplasticity—the brain's ability to reorganize itself by forming new neural connections. This process is crucial for cognitive rehabilitation and can significantly enhance long-term memory and learning adaptation rates.

For parents and caregivers, AI provides invaluable support in monitoring and aiding children's cognitive development. Interactive platforms often include dashboards that track learning milestones, behavioral changes, and cognitive improvements. This form of vigilant yet unobtrusive monitoring allows caregivers to make informed decisions about interventions, educational adjustments, and additional resources to support the cognitive growth of their loved ones.

Moreover, the social implications of enhanced learning and memory through AI extend beyond individual benefits. Empowering individuals with cognitive disabilities to better manage their learning and memory translates to greater independence and active participation in community life. This holistic enhancement permeates various aspects of daily living, be it academic pursuits, professional engagements, or social interactions.

AI's potential in aiding cognitive functions is not without its challenges. Data privacy, ethical considerations, and the digital divide are pressing issues that must be addressed to ensure equitable access to AI-driven learning tools. Safeguarding personal information while providing adaptive learning experiences requires robust security frameworks and ethical guidelines. Similarly, steps must be taken to make these technologies accessible to varied socio-economic demographics to prevent further marginalization.

In conclusion, the integration of AI in improving learning and memory for individuals with cognitive disabilities holds immense promise. By providing personalized, adaptive, and accessible learning environments, AI enables individuals to overcome cognitive barriers, fostering greater independence and enhancing quality of life. The path forward involves continuous innovation, ethical vigilance, and collaborative efforts to ensure that AI's potential is fully realized in the service of those who need it most.

Chapter 9:
AI in Education

The transformative potential of AI in education must not be underestimated; it holds the key to creating personalized learning environments that cater to the specific needs of students with disabilities. By leveraging AI, educational tools can adapt to individual learning styles and pace, making education more accessible and engaging for all. Additionally, AI-driven insights can help bridge educational gaps by providing real-time feedback and tailored support, ensuring that no student is left behind. Innovative technologies such as intelligent tutoring systems and speech recognition software have already shown promise in enhancing inclusivity and fostering an equitable learning experience. By continuously advancing these AI applications, educators can create a more inclusive educational landscape that empowers every student to reach their full potential.

Personalized Learning Environments

When it comes to education, one size rarely fits all. This is especially true for individuals with disabilities, who often face unique learning challenges that standard educational models can't adequately address. With the advent of artificial intelligence (AI), personalized learning environments are becoming increasingly viable, offering customized educational experiences that cater to the specific needs and abilities of each student.

AI has the potential to transform education by creating individualized learning pathways. These systems leverage machine learning algorithms to analyze a student's performance, behaviors, and preferences in real-time, continuously adapting to optimize their learning experience. Whether it's through interactive tutorials, adaptive quizzes, or tailored study plans, AI-powered platforms can provide the necessary academic accommodations without compromising the integrity of the educational content.

Imagine a classroom where every student has an AI tutor tailored specifically for them. For students with disabilities, this could mean the difference between merely getting by and truly excelling. AI can identify the areas where a student struggles and offer additional support in those specific topics. For example, if a student with dyslexia frequently misreads certain words, the system can highlight these patterns and provide exercises that focus on improving reading skills in those areas.

An essential component of personalized learning environments is their ability to offer multiple modes of content delivery. This functionality can be particularly advantageous for individuals who experience impairments in vision or hearing. For instance, a lesson could be presented in text, audio, and video formats, ensuring that the material is accessible regardless of the student's primary mode of learning. This sort of inclusive design reflects a more profound understanding of the diverse ways humans absorb information.

Let's not forget the emotional and psychological aspects of learning. Students with learning disabilities often face stigmatization and lower self-esteem. AI can provide a non-judgmental and patient tutor, which could help in building a student's confidence and making the learning process more enjoyable. The AI can give immediate positive feedback, log progress meticulously, and celebrate small victories along the educational journey—all crucial factors in maintaining motivation and engagement.

Furthermore, AI in personalized learning environments offers valuable data to educators. Through detailed analytics, teachers can gain insights into a student's progress that they might not have been able to discern through traditional methods. This data allows for more informed decision-making, enabling educators to tailor their teaching strategies to better meet the needs of their students. It also streamlines administrative tasks, allowing instructors to focus more on interactive and engaging teaching methods.

However, the implementation of AI in educational settings also requires careful consideration of ethical and privacy concerns. It's crucial to ensure that data is securely stored and that students' privacy is always maintained. Transparency in how data is used and who has access to it is paramount. Striking the right balance between personalized learning benefits and ethical considerations is essential to harnessing the full potential of AI in education.

Additionally, personalized learning environments must be designed with inclusivity at their core. Features such as voice recognition for students with motor disabilities, predictive text suggestions for those with dyslexia, and eye-tracking capabilities for students who can't use traditional input methods can significantly enhance accessibility. These AI-driven tools should not only adapt to a student's learning needs but also support various ways of interacting with the educational content.

The role of parents and caregivers should also not be underestimated. AI systems can provide detailed reports to parents, keeping them informed of their child's progress and areas that need attention. This increased transparency allows for a collaborative approach to education, where parents, teachers, and AI work together to create the most supportive learning environment possible. For many families, this level of insight and involvement is invaluable, providing peace of mind and a clear path forward.

Moreover, personalized learning environments can play a crucial role in lifelong learning. For individuals with disabilities, education doesn't end at the school gates. AI can continue to offer personalized learning opportunities throughout a person's life, facilitating career advancement and personal development. The ability to continuously learn and adapt is invaluable in a rapidly changing job market, and AI can provide the tools necessary to stay competitive.

Another exciting aspect is the potential for peer-to-peer learning facilitated by AI. In these environments, students can collaborate on projects, share knowledge, and support each other in ways that are accessible to everyone. AI can guide these interactions to ensure they are productive and inclusive, breaking down barriers that might otherwise inhibit effective collaboration.

What makes AI-driven personalized learning environments so compelling is their capacity for scalability. Schools across different regions, regardless of their resource levels, can implement these technologies to ensure that all students have access to high-quality education. This democratisation of education can help bridge the gap for students who might not have had access to specialized learning resources otherwise.

Ultimately, the promise of AI in personalized learning environments lies in its ability to make education not just accessible, but also effective and engaging for all students, including those with disabilities. This technology represents more than just an incremental improvement; it stands as a paradigm shift towards truly inclusive education. By leveraging the power of AI, we can create educational experiences that are not just personalized but also transformative, helping every student reach their full potential.

An inclusive education system built on AI doesn't just benefit students with disabilities; it enriches the learning environment for everyone. It fosters a culture of empathy and understanding, teaches the value of diversity, and prepares students for a world where technology and human capabilities are increasingly intertwined. Personalized learning environments are a testament to how AI can be harnessed for the greater good, opening up new pathways to knowledge and personal growth for everyone.

As we continue to develop these technologies, it's essential to remain focused on the broader goal: empowering every learner to succeed. The journey to fully realized personalized learning environments will require ongoing innovation, collaboration, and commitment to ethical practices. But the potential rewards—a more inclusive, accessible, and effective educational experience—are well worth the effort.

Bridging Educational Gaps

Education is more than just a right; it's an essential pillar for personal and societal growth. Yet, for individuals with disabilities, accessing quality education has traditionally been fraught with challenges. The advent of AI technologies offers an unprecedented opportunity to bridge these educational gaps, providing equitable access to learning resources and leveling the playing field.

Consider the nature of personalized learning. Traditional one-size-fits-all approaches often fail to cater to the unique needs of students with disabilities. However, AI can revolutionize this by creating personalized learning experiences tailored to each student's abilities and requirements. Through adaptive learning platforms, AI algorithms assess a student's strengths and weaknesses, customizing content accordingly. This targeted approach not only bolsters educational achievements but also fosters a more inclusive learning environment.

Inclusive Intelligence

Adaptive learning isn't just about modifying content; it's about dynamically adjusting the pace of learning as well. For instance, a student with a cognitive disability might need more time to grasp specific concepts. AI can identify these needs in real-time, providing additional resources or altering the curriculum's difficulty level to prevent the student from feeling overwhelmed. Essentially, AI transforms the educational experience into something fluid and responsive, offering continuous support while respecting individual learning curves.

Language barriers present another significant educational hurdle. For students with hearing impairments, traditional classrooms can be isolating environments. This is where AI-powered speech recognition and conversion tools come into play. These solutions transcribe spoken words to text in real-time, allowing students to follow along without missing a beat. Additionally, AI-driven sign language interpreters can offer instant translations, making lessons universally accessible. Such innovations make it possible for all students to participate fully regardless of their auditory capabilities.

The promise of AI extends to improving educational content accessibility for visually impaired students as well. AI-driven applications can read and interpret text, converting it into auditory or tactile formats. This allows students to consume educational material through whichever medium suits them best. Furthermore, AI tools equipped with natural language processing (NLP) can even summarize complex texts or highlight key sections, enabling visually impaired students to engage with material more interactively and effortlessly.

For physical disabilities, attending traditional classrooms may not always be feasible. AI enhances remote learning possibilities through platforms that facilitate virtual classrooms and interactive experiences. Using virtual reality (VR) and augmented reality (AR), AI can create immersive environments where students feel as though they are part of a physical classroom. This not only mitigates the physical limitations associated with commuting but also enriches the overall educational experience.

Beyond these tailored solutions, AI also plays a pivotal role in addressing the broader issue of resource allocation. Schools and institutions often face constraints on educational resources, which can disproportionately impact students with disabilities. AI can optimize resource management, ensuring that tools, aids, and special services are allocated efficiently according to identified needs. This results in a more equitable educational landscape where every student's requirements are met adequately.

It's also worth noting the significance of AI in assisting educators themselves. Teachers often juggle multiple responsibilities, from lesson planning to student assessments. AI can alleviate some of these burdens by automating administrative tasks and providing data-driven insights into student performance. With AI handling these ancillary tasks, educators can focus more on personalized instruction and support, particularly for students with disabilities.

One of the most impactful ways AI can bridge educational gaps is by fostering collaborative learning. AI-powered platforms can connect students with diverse backgrounds and abilities, encouraging peer-to-peer learning. Such environments promote empathy, understanding, and collective problem-solving, enriching the academic experience for all participants. This creates a culture of inclusivity and mutual support, integral for holistic educational development.

The integration of AI in education also requires a concerted effort toward ethical considerations and inclusivity. AI systems should be designed with the aim of eliminating biases that could further disadvantage students with disabilities. This necessitates collaboration between technologists, educators, and disability advocates to ensure that AI tools are equitable and just. Ethical AI development should focus on transparency, accountability, and inclusivity, making sure that these technologies enhance rather than impede educational equity.

We must recognize that the successful implementation of AI in education involves continuous feedback and adaptation. AI systems should be regularly updated based on user experiences and evolving needs. Engaging students, educators, and parents in this iterative process ensures that the technology remains relevant and beneficial. The commitment to constant improvement is what makes AI a robust tool for bridging educational gaps effectively.

While AI holds enormous potential, it's crucial that we also bridge the digital divide. Access to AI-driven educational tools should not be a privilege but a standard. This means investing in necessary infrastructure, such as high-speed internet and suitable devices, especially in underserved and rural areas. Public policies and private initiatives must collaborate to ensure that the benefits of AI are universally accessible, thereby truly leveling the educational playing field.

One inspiring example is the implementation of AI-driven learning aids in inclusive classrooms across various countries. These initiatives have demonstrated remarkable improvements in both engagement and performance among students with disabilities. By providing personalized instructional support, visual and auditory aids, and fostering collaborative learning environments, AI has proven to be a game-changer in creating an inclusive educational experience.

In conclusion, bridging educational gaps through AI is not just about technological innovation; it's about reimagining education itself. By leveraging AI, we can create dynamic, personalized, and inclusive learning environments. This empowers students with disabilities to achieve their full potential and ensures that no one is left behind. It's a transformative journey that requires commitment, collaboration, and continuous effort, but the rewards—a more inclusive and equitable education system—are worth every step.

Chapter 10: Healthcare Innovations

In the evolving landscape of healthcare, AI-powered innovations are reshaping the way medical professionals diagnose and treat ailments, particularly benefitting those with disabilities. These technological advancements offer unprecedented precision in medical diagnostics, enabling early detection and intervention that can significantly enhance patient outcomes. Personalized healthcare, driven by AI algorithms, tailors treatment plans to suit individual needs, optimizing therapeutic effectiveness and minimizing side effects. Beyond diagnosis and treatment, AI assists in patient management by predicting potential health issues and recommending proactive measures, thereby empowering individuals with disabilities to maintain healthier, more independent lives. Companies and institutions are increasingly adopting these AI solutions, heralding a new era in healthcare that promises greater accessibility and quality of life for all.

AI in Medical Diagnostics

The field of medical diagnostics has witnessed a transformative shift with the incorporation of artificial intelligence (AI). AI technologies are shaping healthcare by providing timely and accurate diagnoses, which is crucial for early intervention and better patient outcomes. The role of AI in this domain is fast evolving, offering a new horizon of possibilities that empowers not only healthcare professionals but also individuals with disabilities who often face additional health challenges.

One of the primary ways AI contributes to medical diagnostics is through image recognition capabilities. Advanced algorithms can analyze medical images such as X-rays, MRIs, and CT scans with a level of precision sometimes surpassing human experts. By identifying patterns and anomalies that might be missed by the human eye, AI systems can provide earlier and more reliable diagnoses. This capability is extraordinarily beneficial for individuals with disabilities, who might require specialized and frequent medical imaging.

Another significant application lies in predictive analytics. AI models can process vast amounts of data to predict the likelihood of diseases, allowing for preventive measures to be taken. For example, AI can analyze electronic health records to forecast potential complications in patients with chronic conditions like diabetes or cardiovascular diseases. By anticipating problems before they become critical, healthcare providers can significantly improve the quality of life for patients, reducing hospital visits and lengthy treatments.

AI-driven tools such as chatbots and virtual assistants also play a crucial role in medical diagnostics. These tools can perform initial assessments based on symptoms reported by patients, offering a preliminary diagnosis or recommending further tests. For people with disabilities, who may have challenges accessing healthcare facilities, these AI tools provide an accessible way to obtain initial medical advice and reduce the strain on healthcare resources.

The integration of AI in medical diagnostics is not just about technical advancements; it also brings an element of personalization. AI systems can tailor their analyses based on an individual's specific health history and conditions. Personalized diagnostic approaches mean that treatment plans can be designed to cater specifically to the patient's unique needs, enhancing the effectiveness of interventions and resulting in better health outcomes.

Moreover, AI algorithms help in the standardization of diagnostic processes. Disparities in healthcare arise partly due to human error and subjective judgments. By standardizing diagnostics, AI ensures that all patients, including those with disabilities, receive consistent and fair evaluations irrespective of their location or the healthcare professional attending to them. This level of uniformity in medical diagnosis is vital for an inclusive healthcare system.

Enhancements in natural language processing (NLP) also pave the way for better diagnostic processes. NLP algorithms can scrutinize doctors' notes, medical literature, and even patient history to identify relevant information that might influence a diagnosis. This comprehensive approach ensures no detail is overlooked, providing a more holistic view of the patient's health status.

Another promising area is the use of AI in genomics. AI-powered tools can analyze genetic data to identify markers related to various diseases, offering a route to genetic-based diagnostics and personalized medicine. This capability is especially relevant for inherited conditions, providing families and individuals with disabilities a clearer picture of their health risks and potential preventative measures.

AI also plays a role in screening for mental health conditions, which can often be overlooked in traditional diagnostic settings. Machine learning models can detect subtle cues in speech, behavior, and social media activity that might indicate mental health issues. For people with disabilities, who might experience higher rates of depression or anxiety, these AI-driven screenings offer a non-intrusive way to monitor mental well-being.

When it comes to implementing AI in medical diagnostics, collaboration between technology developers and healthcare professionals is key. Medical experts provide the domain knowledge necessary to train AI systems effectively, ensuring that they produce meaningful and accurate diagnostic outcomes. This synergy is crucial for developing tools that genuinely supplement human expertise rather than merely replacing it.

Despite the advancements, there are challenges in the integration of AI in medical diagnostics. Issues related to data privacy, security, and ethical considerations must be meticulously addressed to ensure patient trust and the ethical use of AI technologies. Transparent AI systems that can explain their diagnostic processes are essential for gaining the confidence of both healthcare providers and patients.

In conclusion, the future of AI in medical diagnostics looks promising, bringing a new era of precision, personalization, and accessibility. These advancements hold particular importance for individuals with disabilities, who often experience unique healthcare challenges. As AI continues to evolve, its potential to revolutionize medical diagnostics and improve patient care is immense. Embracing these technologies while navigating their ethical landscapes promises a brighter and healthier future for all.

AI for Personalized Healthcare

In the rapidly evolving field of healthcare, the integration of AI has opened unprecedented opportunities, particularly for personalized care. The idea is simple yet groundbreaking: leveraging artificial intelligence to tailor healthcare solutions to an individual's unique needs and circumstances. Personalized healthcare can drastically improve patient outcomes, streamline medical processes, and significantly enhance accessibility for people with disabilities.

First and foremost, AI in personalized healthcare means customized treatment plans. Traditional treatment approaches often follow a one-size-fits-all model, which may not consider the unique physiological and psychological makeup of each patient. AI algorithms, however, can sift through vast amounts of data—including genetic information, lifestyle choices, and previous medical histories—to find the most effective treatment options. This form of precision medicine is not only more effective but often less invasive and time-consuming.

One of the key components in personalized healthcare is predictive analytics. Machine learning models can predict the likelihood of disease development based on various data points. This kind of foresight is invaluable for individuals with disabilities who may already be managing chronic conditions. Early intervention can mean the difference between a manageable condition and significant deterioration in quality of life. For instance, AI can predict flare-ups in chronic illnesses like Multiple Sclerosis, allowing for timely preventative measures.

Another crucial aspect is patient monitoring. AI-powered wearables and home monitoring systems collect real-time data, providing continuous health updates. This is especially beneficial for those with mobility impairments or severe disabilities that make frequent doctor visits challenging. Wearables can monitor vitals like heart rate, blood pressure, and oxygen levels. The AI analyzes these data points and can alert healthcare providers to any alarming changes. This kind of monitoring is not just convenient; it can be life-saving.

Moreover, personalized healthcare solutions are also extending into the realm of mental health. AI chatbots and virtual therapists can offer immediate, personalized emotional support tailored to an individual's psychological profile. These tools are particularly beneficial for individuals with cognitive disabilities, who may find traditional therapy challenging. Machine learning models can adapt based on user interactions, making each session more personalized and effective over time.

Diagnostics have also seen significant advancements thanks to AI. Machine learning algorithms can recognize patterns in medical imaging and lab results with a high degree of accuracy. For individuals with disabilities, this means quicker, more accurate diagnoses and less strain from prolonged testing periods. For example, AI software that analyzes retinal images can detect diabetic retinopathy with greater speed and accuracy compared to traditional methods, thereby preventing potential blindness through early intervention.

Another aspect where AI shines is medication management. For many people with disabilities, managing multiple medications can be a daunting task. AI systems can recommend optimal medication schedules, dosage adjustments based on real-time monitoring, and even predict potential adverse drug interactions. This kind of precision ensures that patients are receiving the most effective treatment with the least risk of side effects.

Rehabilitation and physical therapy are also benefiting from AI. Personalized rehabilitation programs using robots and AI-driven exercise systems can adapt to the individual progress of patients with physical disabilities. These systems can provide real-time feedback and adjustments to exercise routines, making sessions more effective and engaging. AI can even gamify these programs, increasing motivation and adherence.

Importantly, we've also seen a significant push towards making healthcare more inclusive through AI. Traditional healthcare systems can inadvertently exclude various groups, especially those with disabilities, due to inherent biases or accessibility issues. AI can help identify and mitigate these biases. For instance, machine learning models can be trained on diverse datasets to ensure they provide accurate recommendations across different demographics, including those with disabilities.

The ethical considerations surrounding AI in healthcare can't be overlooked. Privacy concerns, data security, and algorithmic transparency are critical issues that need to be meticulously addressed. People with disabilities often face additional vulnerabilities, making it imperative that AI solutions are both secure and ethical. Healthcare providers must ensure that data is anonymized and stored securely, and that AI recommendations are transparent and interpretable.

Lastly, the financial implications should not be ignored. Implementing AI in personalized healthcare does come at a cost, but it's an investment that pays dividends in the long run. Reduced hospital admissions, more effective treatments, and quicker recovery times can result in substantial healthcare savings. For individuals with disabilities, this means more affordable and accessible care, ultimately leading to an improved quality of life.

In conclusion, AI for personalized healthcare represents a transformative shift in the way we approach medical care. By tailoring treatments, improving diagnostics, and providing continuous monitoring, AI can significantly enhance the lives of individuals with disabilities. However, to realize its full potential, it's crucial to address the ethical and financial aspects diligently. As AI continues to evolve, the promise of a more accessible, efficient, and personalized healthcare system becomes increasingly attainable, offering new hope and possibilities for millions around the globe.

Chapter 11:
Mental Health and Emotional Support

Mental health is a vital yet often overlooked aspect of well-being, especially within the disability community. AI technologies have the potential to revolutionize emotional support by providing accessible, personalized, and immediate interventions. From AI-driven mental health monitoring tools that can detect subtle changes in mood, behaviors, or speech, to interactive platforms offering therapeutic conversations and coping strategies, technology can offer significant assistance. These tools enable continuous and objective mental health tracking, which can be crucial for early intervention and long-term support. The integration of AI in emotional wellbeing tools, like virtual companions and empathetic chatbots, allows for companionship and a sense of connection, which are indispensable in combating isolation and loneliness. By bridging gaps in traditional mental health services, AI not only enhances care but also restores independence and dignity for individuals with disabilities, fostering a more inclusive and supportive environment for all.

AI for Mental Health Monitoring

The integration of artificial intelligence (AI) into mental health monitoring holds incredible promise for transforming the lives of individuals facing mental health challenges. At the core, AI is enabling more timely, personalized, and effective interventions, which can be crucial in managing mental health conditions. The power of AI in this realm stems from its ability to process vast amounts of data swiftly, identify patterns that might be missed by human observers, and provide real-time feedback and support.

One of the most significant advantages of AI-driven mental health monitoring is its capability for early detection. AI systems can analyze data from various sources such as social media, wearable devices, and electronic health records. By evaluating changes in behavior, language, and physiological signals, AI can flag potential issues before they escalate into severe conditions. This proactive approach allows for early intervention, which is often pivotal in mental health care.

Consider, for example, an individual who frequently uses a mental health app that tracks mood and activity levels. The AI within the app might notice a consistent decline in mood and daily activities, signaling a potential depressive episode. The system could then alert the user and suggest reaching out to a mental health professional, providing a lifeline at a critical moment. This constant vigilance and early warning system offer users a safety net that previously did not exist.

Moreover, AI doesn't just stop at detection; it also plays a substantial role in ongoing support and management of mental health. Chatbots and virtual therapists, powered by natural language processing and machine learning, are becoming increasingly sophisticated. These AI-driven tools can offer emotional support, cognitive behavioral therapy (CBT) techniques, and other therapeutic interventions through a smartphone or computer. This accessibility is vital, especially for those who may feel stigmatized or uncomfortable seeking in-person help.

AI-driven mental health tools are designed to be accessible 24/7, breaking down the barriers of time and place. This is particularly beneficial for individuals living in remote areas with limited access to mental health services. It also allows for continuous care and support, which is crucial for managing chronic mental health conditions. For many, just knowing that there is an ever-present, non-judgmental entity to talk to can significantly reduce feelings of isolation and anxiety.

In addition to providing direct support, AI helps in personalizing mental health care. Machine learning algorithms can analyze an individual's responses and behaviors to customize interventions. For instance, if a user finds more relief in mindfulness exercises rather than journaling, the AI system can adapt and offer more of what's beneficial. This personalization ensures that the care provided is not one-size-fits-all but tailored to individual needs, making it more effective.

Another critical aspect of AI in mental health monitoring is data analysis. Mental health professionals traditionally rely on self-reported data and clinical observations, which can be limited and subjective. AI, however, can integrate and analyze large datasets from diverse sources, providing a more comprehensive picture of a person's mental health. Wearable devices, for example, can monitor physiological indicators such as heart rate variability and sleep patterns, offering insights into the biological aspects of mental health conditions.

These AI systems can also facilitate better communication between patients and healthcare providers. By continuously monitoring and analyzing data, AI can generate detailed reports that can be shared with professionals. This assists in making informed decisions regarding treatment plans. It also ensures that healthcare providers are aware of any significant changes in a patient's condition promptly, allowing for quicker adjustments in care.

Nevertheless, the use of AI in mental health monitoring isn't without challenges. Privacy is a major concern due to the sensitive nature of mental health data. Ensuring that data is kept secure and used ethically is essential to gaining the trust of users. Furthermore, while AI can provide valuable support, it shouldn't replace human interaction and professional judgment. The best approach is a hybrid one where AI augments human care rather than substitutes it.

Accessibility and inclusiveness are also areas that need attention. Not everyone has access to the latest technology or an internet connection, which can limit the reach of AI-based mental health solutions. Efforts should be made to bridge this digital divide to ensure that benefits are available to all, irrespective of socioeconomic status or geographic location.

Equally important is the need for ongoing validation and improvement of AI models. Mental health is an evolving field, and AI systems need to adapt continuously to the latest research and developments. This requires collaboration between AI developers, mental health professionals, and end-users to keep the technology relevant and effective.

In the long run, the potential of AI for mental health monitoring is immense. As algorithms become more sophisticated and datasets more comprehensive, AI will likely offer even more precise predictions and customized care plans. This progress will contribute significantly to reducing the global burden of mental health disorders, enhancing the overall quality of life for many individuals.

Ultimately, incorporating AI into mental health monitoring represents a paradigm shift in how we view and manage mental health. It moves us from reactive to proactive care, from one-size-fits-all to highly personalized interventions, and from episodic check-ins to continuous support. By harnessing the power of AI, we can ensure that individuals facing mental health challenges are not alone in their journey and have the tools they need to thrive.

Emotional Wellbeing Tools

In an increasingly digital world, the advent of AI-powered emotional wellbeing tools is revolutionizing mental health and emotional support. Imagine a world where AI companions can detect subtle shifts in your mood, offering timely support and guidance individualized for your emotional needs. This fusion of technology and empathy is not merely a futuristic endeavor; it is an evolving reality poised to enhance emotional wellbeing for individuals with disabilities.

AI can analyze patterns in speech, text, and even facial expressions to understand emotional states. Emotional wellbeing tools leverage advanced machine learning algorithms to identify signs of stress, anxiety, or depression in real-time. For instance, these tools can monitor voice tones during phone calls or analyze text inputs like social media posts and text messages. By doing so, they provide invaluable insights that can inform interventions and offer solace to individuals grappling with mental health challenges.

The applications of these AI tools are multifaceted. They serve as emotional checkpoints, tracking mood fluctuations and offering therapeutic activities based on individual preferences. For example, some AI-driven apps provide mindfulness exercises, guided meditations, or even simple breathing techniques to help users regain emotional balance. The ability to provide personalized interventions makes these tools exceptionally beneficial for those with cognitive disabilities who may struggle to verbalize their emotional states.

Beyond monitoring and intervention, AI emotional wellbeing tools often come equipped with features that promote ongoing emotional growth. Take, for instance, journaling applications augmented by AI. These platforms not only offer a space for self-reflection but also provide analytical feedback on emotional patterns and suggest coping mechanisms. By fostering a better understanding of one's emotional landscape, these tools empower users to take proactive steps in managing their mental health.

The integration of AI into therapeutic practices has also taken significant strides. Virtual therapists driven by AI are continually improving in their ability to deliver cognitive-behavioral therapy (CBT). These virtual therapists can conduct sessions that adapt dynamically to the user's responses, providing tailored treatment plans that evolve in real-time. While they do not replace human therapists, they offer a substantial supplement, particularly for those who might find it difficult to access traditional forms of therapy due to geographical or financial constraints.

For those with physical disabilities, emotional wellbeing tools often integrate seamlessly with other assistive technologies. Consider a smart home ecosystem where AI ensures the entire environment is tuned to support emotional health. From lighting adjustments that create calming atmospheres to music that influences a positive mood, the possible environments are diverse and customizable. This level of integration provides a comprehensive approach to emotional wellbeing, making daily life not only manageable but also enriching.

In educational settings, the use of AI to support emotional wellbeing is growing. These tools can be invaluable for both students and educators in managing stress and improving overall mental health. By recognizing stress indicators, AI systems can prompt supportive interventions that range from suggesting a break to offering motivational messages. Emotional wellbeing tools in educational environments not only benefit students with disabilities but also create a more inclusive and supportive atmosphere for all.

An essential aspect of these tools is their adaptability. The power of AI lies in its continual learning process. Emotional wellbeing tools learn from user interactions, refining their algorithms to better serve individual needs over time. This adaptability ensures that the support provided remains relevant and effective, fostering continuous emotional growth and resilience.

However, it's crucial to recognize that with the deployment of these tools, privacy and ethical considerations come to the fore. Ensuring the secure handling of sensitive emotional data requires stringent data protection measures. Users must feel confident that their personal information is secure and that AI algorithms are transparently designed to avoid biases that could skew emotional assessments. Ethical AI design underscores the importance of fairness and accuracy, fostering trust and reliability in emotional wellbeing tools.

The potential for community building through AI emotional wellbeing tools is another exciting frontier. These tools can facilitate the creation of virtual support groups, matching individuals experiencing similar emotional challenges. These AI-curated groups can provide a sense of community and shared understanding, which is essential for emotional wellbeing. By connecting people who might otherwise feel isolated, AI helps to build supportive networks that offer both virtual and human connections.

Moreover, advancements in natural language processing (NLP) have made it possible for AI to understand and respond to natural human language. This capability allows for more conversational and relatable interactions, making the users feel understood and supported. It is particularly beneficial for individuals with disabilities who may find it challenging to communicate their emotions traditionally. The goal is to create a symbiotic relationship with AI, where users feel heard, understood, and supported in their emotional journeys.

As these tools continue to evolve, the future holds even greater promise. Emerging technologies such as affective computing aim to further bridge the gap between human emotions and machine responses. Affective computing focuses on developing systems that recognize, interpret, and simulate human emotions. The application of this technology could further enhance the capability of emotional wellbeing tools, making them more intuitive and responsive to human needs.

The journey toward integrating AI into emotional wellbeing is a testament to technology's potential to enhance human life. These tools are not just about managing mental health but are designed to enrich emotional experiences and improve quality of life. As we continue to innovate and iterate on these technologies, the ultimate aim is to create inclusive, supportive environments where individuals with disabilities can thrive emotionally and mentally.

In conclusion, the union of AI and emotional wellbeing tools is a burgeoning field that holds immense potential. By offering real-time monitoring, personalized interventions, therapeutic support, and community-building opportunities, these tools are set to revolutionize mental health support for individuals with disabilities. As they continue to evolve, they promise to provide not only valuable support but also a new level of emotional resilience and empowerment.

Emotional wellbeing is a fundamental aspect of human life. Enhancing this through AI-driven tools could indeed be one of the most transformative contributions of our time, helping to create a world where everyone, regardless of ability, can access the support they need to lead emotionally fulfilling lives.

Chapter 12:
Enhancing Workplace Accessibility

Embracing AI-driven solutions in the workplace isn't just an innovative approach—it's a transformative one, especially for employees with disabilities. AI technologies are breaking down barriers, offering unprecedented levels of accessibility, and fostering inclusive environments where everyone can thrive. From AI-powered tools that enhance productivity to remote work applications that ensure flexibility, these advancements are providing individuals with disabilities the means to fully participate in the workforce. Whether it's voice recognition software that offers seamless communication or advanced data interpretation tools that support complex tasks, the integration of AI into the workplace is not only reshaping job functions but also empowering individuals by providing equal opportunities. Let's explore how these technologies are revolutionizing workspaces and unlocking potential, making accessibility not just a compliance checkbox, but a cornerstone of modern workplaces.

AI in the Workplace

Artificial Intelligence (AI) is transforming workplaces by making them more accessible, particularly for people with disabilities. This innovation goes beyond simple automation and touches every aspect of employment, from recruitment and training to daily operations and career advancement. AI technologies offer unique solutions tailored to individual needs, ensuring that the workplace becomes a more inclusive environment.

Consider recruitment. Traditional hiring processes have often been impediments to people with disabilities, who face a myriad of challenges such as non-inclusive interview formats and physical accessibility issues. AI-driven recruitment platforms use algorithms that focus on skill sets rather than physical presence. Tools like video interviewing software with AI capabilities can analyze candidates based on performance metrics void of bias, thus promoting fairness and equal opportunity.

Training and development also benefit significantly from AI. Adaptive learning platforms can provide personalized training modules that cater to different learning styles and speeds. For example, employees with cognitive disabilities can engage with interactive content adjusted to their pace and understanding. This customization enriches the learning experience and ensures that no one is left behind in skills development.

Daily operations in many workplaces are increasingly supported by AI tools. Take communication, for instance. AI-driven speech-to-text applications and real-time translation services have made it possible for employees with hearing or speech impairments to participate fully in meetings and collaborative projects. These tools not only facilitate communication but also foster an inclusive culture where everyone's input is valued.

AI isn't just about software; it's about creating a holistic environment. Physical spaces are also becoming smarter and more accommodating. For people with mobility issues, AI-powered robotic assistants and smart prosthetics offer new levels of independence. Office layouts are being designed to be more navigable, thanks to sensors and AI analytics that assess and optimize spatial configurations.

These physical innovations make workplaces safer and more comfortable for everyone.

Remote work options have seen massive improvements, largely due to AI. Technologies such as virtual reality (VR) meetings and AI-backed project management tools ensure that remote employees, including those with disabilities, don't miss out on collaboration or networking opportunities. AI can facilitate goal setting, task assignment, and progress tracking, making remote work seamless and efficient.

It's not just about easing existing challenges; AI opens up entirely new career paths. For example, an individual with a visual impairment might traditionally struggle in roles that require extensive data analysis. However, with AI tools that provide auditory data readouts or tactile interfaces, these barriers are eroded, allowing access to careers that were previously considered unattainable.

The empowerment doesn't stop with individuals; it extends to teams as well. AI can analyze team dynamics and suggest ways to optimize collaboration. It can track and predict team performance, offering insights on how to leverage each member's strengths and address weaknesses. This level of understanding and optimization ensures that teams operate at their highest potential, benefiting the whole organization.

We must highlight the psychological impact of AI in the workplace. It's not just about enabling tasks; it's also about boosting confidence and reducing the stigma associated with disabilities. When AI tools enable employees with disabilities to perform on par with their peers, it not only proves their capabilities but also enhances their self-esteem and job satisfaction.

The ethical considerations are equally important. While AI has the potential to amplify abilities, inclusive design principles must guide its implementation. Developers and employers need to collaborate with people with disabilities to create AI solutions that genuinely meet their needs. This involves continuous feedback loops and iterative improvements to ensure the technology is truly beneficial.

Furthermore, ongoing education and awareness are critical. Employers and employees alike need to be informed about the possibilities and limitations of AI. Regular training sessions can help demystify AI and present it as a tool for inclusivity rather than as a complex, inaccessible technology. This fosters a more accepting and proactive organizational culture.

Looking ahead, the convergence of AI with other emerging technologies promises even more revolutionary changes. For instance, the integration of AI with Internet of Things (IoT) devices can create interconnected environments that adapt in real-time to the needs of employees with disabilities. Imagine a smart office where lighting, temperature, and layouts adjust automatically according to the user's preferences and requirements.

On a policy level, governments and organizations must push for regulations that encourage the adoption of AI-driven accessible technologies. Incentives, grants, and public-private partnerships can catalyze innovation and deployment. Ethical guidelines and standards must be established to ensure that AI solutions do not inadvertently exclude or disadvantage any group.

The journey towards AI-enhanced workplace accessibility is an ongoing one. It requires dedication, innovation, and, above all, a commitment to inclusivity. However, the benefits far outweigh the challenges. AI is not just a technological advancement; it's a social one. By harnessing its power, we're not only making workplaces better for people with disabilities but also enriching the work environment for everyone.

As we march forward, it's clear that AI will continue to play a pivotal role in shaping accessible workplaces. It's a collaborative effort, pulling in insights from diverse fields to create solutions that are dynamic, effective, and deeply human-centric. The potential to revolutionize the workplace and make it truly inclusive stands as one of the most promising aspects of our AI-driven future.

Tools for Remote Work and Collaboration

In today's interconnected world, remote work is no longer a luxury but a necessity, especially for individuals with disabilities. The shift towards a more flexible workplace has opened doors to numerous opportunities, but it also brings its own set of challenges. Fortunately, AI-powered tools have emerged as game-changers in making remote work more accessible, efficient, and inclusive.

At the heart of these innovations lies the ability to communicate effectively. Tools like AI-driven speech-to-text and text-to-speech programs have dramatically improved the way people with disabilities interact online. For instance, individuals with hearing impairments can now actively participate in virtual meetings through real-time captioning. On the flip side, those with visual impairments benefit from screen readers that can transcribe textual information into spoken words, making digital content more accessible.

Collaboration platforms such as Zoom, Microsoft Teams, and Slack have integrated AI technologies to enhance usability. Real-time translation services break down language barriers, ensuring that team members can collaborate seamlessly regardless of their native language. Moreover, these platforms often include customizable interfaces, allowing users to adapt the software to their specific needs.

AI can also automate routine tasks, mitigating the cognitive load on users. Automated scheduling assistants can coordinate meeting times that suit everyone's calendar, reducing the need for constant back-and-forth communication. These AI assistants can also send reminders, manage emails, and even draft responses – significantly boosting productivity without overburdening the user.

Machine learning algorithms have enabled detailed data analysis and reporting, allowing team members to focus on strategic tasks rather than getting bogged down in the minutiae. Whether it's generating financial reports or analyzing project workflows, AI tools can handle these complex tasks with greater accuracy and speed, benefiting everyone on the team, particularly those with cognitive disabilities.

Virtual reality (VR) and augmented reality (AR) technologies are also stepping up to make remote collaboration more engaging and inclusive. These platforms can create immersive meeting environments, provide hands-on training, and offer virtual office spaces that simulate a traditional workplace. Such technologies are particularly beneficial for individuals with mobility impairments, who might find traveling to a physical office challenging.

Moreover, AI-powered project management tools like Trello and Asana now offer features that can be customized for accessibility. From voice commands to visual alerts, these tools ensure that all team members can contribute effectively, no matter their physical or cognitive limitations. For example, color-coding tasks can benefit individuals with learning disabilities by breaking complex projects into manageable steps.

The rise of remote work has also brought renewed attention to mental health—an area where AI tools are making significant strides. AI-driven mental health apps provide real-time mood tracking, cognitive behavioral therapy exercises, and even virtual counseling sessions. This support is crucial for individuals facing the dual challenges of remote work and managing their disabilities.

It's essential to underscore the importance of personalized experiences in remote work environments. AI systems can adapt to individual preferences, ensuring that the user interface is as comfortable and efficient as possible. From adjustable font sizes to voice-activated commands, these customizations make remote work more manageable and enjoyable for everyone involved.

AI also plays a critical role in enhancing security and privacy in remote work settings. Advanced encryption algorithms and biometric verification methods ensure that sensitive information remains protected, enabling people with disabilities to work with confidence and peace of mind. This level of security is particularly vital for those handling personal health information or other confidential data.

Another promising area is the development of AI-driven gamification strategies to boost engagement and inclusivity. Integrating game elements into work tasks can make daily activities more engaging and less monotonous. For individuals with cognitive disabilities, these strategies can enhance focus, improve task completion rates, and foster a sense of achievement and motivation.

Team-building is also an area that benefits from AI-facilitated remote work tools. AI algorithms can analyze team dynamics and suggest exercises and activities that promote cohesion and trust. Virtual escape rooms, trivia games, and other interactive sessions can be tailored to ensure inclusivity, making everyone feel an integral part of the team.

Furthermore, AI's role in remote training and skill development cannot be overstated. Adaptive learning platforms that adjust to an individual's pace and learning style make it easier for employees to acquire new skills. For people with disabilities, this means access to tailored educational content that accommodates their unique needs, whether through simplified text, visual aids, or interactive simulations.

The power of AI to level the playing field in remote work extends to performance evaluation and career growth as well. Bias in performance reviews can be mitigated through AI's objective analysis of metrics and achievements. This leads to fairer evaluations, promotions, and opportunities for individuals with disabilities, helping them progress in their careers based on merit.

The synergy of AI and remote work tools isn't just making the workplace more accessible; it's transforming the very nature of work itself. By harnessing the capabilities of these technologies, organizations can create a more inclusive and effective workforce. As AI continues to advance, the potential for even greater inclusivity and productivity is bound to expand, offering unprecedented opportunities for individuals with disabilities.

Chapter 13:
AI in Communication Aid

AI in communication aid is revolutionizing how individuals with speech and language impairments interact with the world. By leveraging advanced algorithms and machine learning, AI-powered communication devices now offer seamless, intuitive interfaces that translate thoughts into speech, making real-time conversation more natural and fluid. Natural language processing (NLP) applications have opened up new avenues for those who previously faced insurmountable barriers in expressing themselves. From predictive text technologies to voice recognition systems, AI is breaking down communication barriers and fostering independence. These innovations don't just offer practical benefits; they restore dignity and confidence, ensuring that everyone has a voice in our increasingly interconnected society.

Advanced Communication Devices

The convergence of AI technology with communication aids has revolutionized how individuals with disabilities interact with the world. Advanced communication devices now offer more than just the ability to express basic needs; they empower users to engage in complex conversations, participate in social interactions, and live more independently. These innovations incorporate sophisticated algorithms, machine learning, and natural language processing to deliver seamless and intuitive communication experiences.

One significant breakthrough in advanced communication devices is the development of AI-driven speech-generating devices (SGDs). These devices are designed to aid individuals with speech impairments, providing them with a voice to express themselves. Powered by AI, modern SGDs can predict words and phrases based on context, improving communication speed and accuracy. For example, rather than typing out every word, users can rely on predictive text features to facilitate quicker and more fluid conversations.

Moreover, AI-enhanced SGDs can be tailored to individual speech patterns and preferences, offering a personalized communication experience. For instance, by analyzing a user's distinct cadence and commonly used expressions, these devices can create more natural and relatable speech outputs. The integration of voice synthesis technology further enriches this experience by allowing users to choose or even create a voice that resonates with their identity, contributing to a more authentic and dignified mode of communication.

In addition, advancements in sensor technologies have opened new avenues for communication aids. Eye-tracking systems, for instance, have become a game-changer for individuals with severe physical disabilities. These systems, combined with AI algorithms, enable users to control communication devices using only their eye movements. By gauging where the user is looking, the device interprets their intentions and translates them into spoken or written words. This not only restores their ability to communicate but also enhances their autonomy and quality of life.

Gesture recognition technology has also gained traction in the realm of advanced communication devices. Utilizing machine learning models, these systems can interpret and translate sign language or other gestures into verbal language. This presents a remarkable opportunity for individuals who primarily communicate through sign language to engage with those who do not understand it, bridging the communication gap and fostering inclusivity.

Another noteworthy innovation in this domain is the advent of brain-computer interfaces (BCIs). These devices harness the power of AI to interpret brain signals and convert them into actionable commands. While still in the experimental phase, BCIs hold promise for individuals with locked-in syndrome or other severe motor disabilities, offering them a way to communicate directly through their thoughts. As research and development continue, we can anticipate more refined and accessible BCI-based communication aids becoming available.

Wearable technology provides yet another dimension to advanced communication solutions. Smartwatches and other wearable devices equipped with AI capabilities can offer real-time translation services, converting spoken language into text and vice versa. This is immensely beneficial for individuals with hearing impairments or those who face language barriers. The ability to access immediate translations facilitates smoother interactions and enhances participation in diverse settings.

AI is also transforming traditional word processing software into intelligent communication tools. Text prediction, autocorrect, and speech-to-text functionalities, powered by neural networks, enable users to compose messages with minimal effort. These features are especially useful for individuals with dexterity challenges, allowing them to communicate more efficiently on digital platforms such as email, social media, or online forums.

Beyond individual devices, the connectivity of communication aids to the Internet of Things (IoT) has broadened their utility. AI-equipped communication devices can now interact with smart home systems, enabling users to control various aspects of their environment through simple commands. This capability extends communication aids from mere speech tools to comprehensive control hubs that enhance independence and usability, especially for individuals with mobility impairments.

Voice assistants, such as Amazon's Alexa and Google Assistant, have become ubiquitous and are increasingly integrated into communication aids. These assistants use natural language processing (NLP) to understand user commands and facilitate tasks ranging from setting reminders to managing daily schedules. The integration of voice assistants into communication aids can ease various daily tasks, thus enhancing the user's independence and quality of life.

One story illustrative of these innovations is that of Sarah, a young woman with cerebral palsy. Her advanced communication device, a combination of eye-tracking technology and NLP, has transformed her ability to communicate. What once took her several painstaking minutes to type out on a keyboard can now be expressed in seconds. She can converse more naturally with friends and family, participate actively in her studies, and pursue hobbies such as writing poetry. The device's AI capabilities have not only given Sarah a voice but also a robust platform for self-expression and personal growth.

Educational institutions are also benefitting from advanced communication aids. Tools that assist students with disabilities ensure that they can engage fully in classroom discussions and activities. AI-powered communication devices can translate classroom lectures in real-time, provide personalized learning aids, and even support collaborative projects by facilitating better communication among peers.

Healthcare professionals are finding advanced communication devices indispensable in patient care. These tools empower patients to express their symptoms, needs, and concerns accurately, leading to more effective and personalized treatment plans. For practitioners, it means less guesswork and miscommunication, fostering a more collaborative and efficient healthcare environment.

While the advancements are significant, it is important to recognize and address potential challenges. Accessibility, affordability, and user-friendliness remain top priorities. Developers must ensure that advanced communication devices are inclusive by design, catering to diverse needs and skill levels. Moreover, ethical considerations, such as data privacy and consent, need to be precisely managed to build trust and ensure the safety of users.

Collaborations between technology companies, healthcare providers, academic researchers, and end-users are crucial for the continuous improvement of these communication aids. By actively involving users in the development process, innovators can create AI-driven solutions that are not only cutting-edge but also deeply aligned with the needs and expectations of those they aim to serve.

The future of advanced communication devices is both exciting and promising. As AI technology continues to evolve, we can anticipate even more intuitive, responsive, and accessible tools that will further enrich the lives of individuals with disabilities. This ongoing innovation underscores a motivational journey towards breaking communication barriers, fostering equality, and creating a world where everyone has the opportunity to voice their thoughts and connect with others.

Natural Language Processing Applications

Natural Language Processing (NLP) is a branch of artificial intelligence that focuses on the interaction between computers and humans through natural language. NLP allows machines to understand, interpret, and generate human language in a valuable way. For individuals with disabilities, especially those with communication impairments, NLP applications have become a transformative tool in breaking down barriers and enhancing daily life.

One of the most impactful applications of NLP in communication aid is through text-to-speech and speech-to-text technologies. For individuals who have difficulties speaking, text-to-speech systems can convert written text input into spoken words, providing a voice to those who might otherwise remain silent. For those who struggle with writing or typing, speech-to-text tools can transcribe spoken words into text, facilitating communication across multiple platforms, such as emails, messaging apps, and social media. These technologies are not only useful for personal communication but also play a critical role in professional and educational settings, ensuring that individuals can participate fully and effectively.

Moreover, NLP enhances the functionality of augmented and alternative communication (AAC) devices. These devices, which range from basic picture boards to sophisticated digital systems, help individuals with speech and language disabilities communicate more easily. Advanced NLP algorithms allow AAC devices to predict words and phrases based on user input, making communication more intuitive and efficient. For example, predictive text and autocorrect functionalities can significantly speed up the communication process, reducing the time and effort required to form sentences.

Virtual assistants, powered by NLP, also represent a significant advancement for people with disabilities. Tools like Amazon's Alexa, Apple's Siri, and Google Assistant can carry out a variety of tasks via voice commands, from setting reminders and sending messages to controlling smart home devices. For individuals with mobility impairments or visual disabilities, these virtual assistants can provide greater independence and control over their environment. By using simple verbal commands, users can perform tasks that might otherwise require physical assistance, enhancing autonomy and quality of life.

NLP applications extend to real-time translation services, which are particularly beneficial for those communicating non-verbally or in multiple languages. Services like Google Translate utilize NLP to convert spoken or written text from one language to another in real time. This capability is invaluable in multicultural environments where language barriers might otherwise impede effective communication. For instance, a student with hearing impairment can use translation tools to follow along in a classroom where the instruction is in a different language, thus accessing educational resources that were previously out of reach.

Another exciting NLP application is sentiment analysis, which can assess and interpret the emotional tone behind a string of text. This technology can be particularly useful for identifying emotional states in written communication, enabling caregivers and healthcare providers to better understand and respond to the needs of individuals with disabilities. For instance, text messages or emails analyzed for emotional content can alert caregivers to potential distress or emotional issues, allowing for timely intervention and support.

Real-time captioning is another area where NLP has made significant strides. Automatic speech recognition (ASR) systems convert spoken language into written text instantaneously, providing real-time captions for live events, video calls, webinars, and more. These services are invaluable for individuals who are deaf or hard of hearing. By providing accurate and timely subtitles, these systems ensure that users can fully engage with audiovisual content, both in professional settings and in their personal lives. Recent advancements in machine learning and NLP have improved the accuracy and reliability of real-time captioning, making it a viable solution for accessibility in various contexts.

Additionally, NLP is increasingly being integrated into social media platforms to assist users with disabilities. Features such as automated image descriptions and alt text generation help individuals with visual impairments navigate and participate in social media more effectively. By analyzing the content of images and generating descriptive text, NLP algorithms provide context and detail that would otherwise be inaccessible. This not only makes the online world more inclusive but also encourages social interaction and community building.

Beyond direct communication tools, NLP also enhances educational technologies that support learning and development for individuals with cognitive disabilities. For example, intelligent tutoring systems leverage NLP to provide personalized feedback and instruction based on the user's input. By understanding and responding to natural language queries, these systems can offer tailored educational experiences that accommodate unique learning styles and needs. This level of personalization helps bridge educational gaps and ensures that all learners have access to the resources they need to succeed.

The integration of NLP in healthcare communication is another noteworthy application. Digital health tools utilizing NLP can assist in patient-provider interactions, particularly for patients with speech or language impairments. These tools can transcribe patient speech into medical notes or translate medical information into simpler language, ensuring that patients understand their health conditions and treatment plans. Furthermore, NLP can be used to analyze patient records and extract relevant information quickly, facilitating more efficient and effective healthcare delivery.

Innovation in NLP doesn't stop at these current applications. Researchers and developers are continually exploring new ways to leverage natural language understanding to improve communication aids. For instance, there is ongoing work in developing NLP systems that can detect and adapt to regional dialects and accents, making technology more accessible to a diverse user base. Similarly, efforts are being made to create more context-aware and empathetic virtual assistants that can understand and respond to the subtleties of human emotion and intent more accurately.

In conclusion, the applications of Natural Language Processing in communication aids are vast and varied, offering significant benefits to people with disabilities. By enhancing text-to-speech and speech-to-text technologies, improving AAC devices, enabling real-time translation and captioning, and supporting personalized educational tools, NLP is breaking down communication barriers and opening up new possibilities for independence and inclusion. As AI technology continues to evolve, so too will the potential for NLP to transform the lives of individuals with disabilities, offering them greater agency and a louder voice in the world.

Natural Language Processing, with its dynamic capabilities, is not just a technological advancement but a beacon of hope and empowerment. It illustrates the profound impact that well-designed AI solutions can have in fostering a more inclusive society where every individual, regardless of their abilities, can communicate and connect more easily. The promise of NLP lies in its ability to continually adapt and improve, reflecting the diverse and intricate needs of its users, and paving the way for a future where communication is truly accessible for all.

Chapter 14:
Legal and Ethical Considerations

As artificial intelligence continues to shape the future of accessibility and independence for individuals with disabilities, understanding the legal and ethical landscape becomes crucial. Addressing privacy concerns is paramount—ensuring data collected by AI systems is secure and utilized responsibly can maintain trust and dignity for users. Equally important is the development of fair and inclusive AI to prevent biases that could amplify existing disparities, making technology truly accessible for everyone. Navigating these considerations requires collaborative efforts from policymakers, developers, and end-users to create robust frameworks that uphold human rights while fostering innovation. Only by achieving this balance can AI reach its full potential as a transformative ally for people with disabilities.

Privacy Concerns

When we talk about the intersection of artificial intelligence and disability, it's impossible to ignore the substantial privacy concerns that come along with it. At the core, these concerns revolve around the collection, storage, and use of sensitive personal data. AI systems, particularly those designed to assist individuals with disabilities, often require access to very private information. This could include medical histories, daily routines, and even real-time tracking data. While the benefits of AI in enhancing accessibility and independence are manifold, they must be balanced with robust privacy protections to ensure that the users' rights and dignity are upheld.

One of the primary concerns is the data collection process itself. To function effectively, AI systems need vast amounts of data, which often means continuous monitoring and recording of users' behaviors and interactions. For instance, smart assistants and AI-powered personal aides designed to help with daily living tasks may need to track a user's location, listen to conversations, or even record video footage. While these features provide invaluable support, they also raise significant privacy issues. Users must have transparent and overt control over their data, knowing exactly what is collected, how it is stored, and for what purposes it will be used.

The storage of data presents another layer of concern. Storing sensitive information online or in cloud-based systems exposes it to potential cyberattacks and unauthorized access. Data breaches can have dire consequences, particularly for individuals with disabilities who might rely on these AI systems for critical daily functions. Robust encryption methods and stringent access controls are essential to safeguard this sensitive information. Furthermore, developers need to be transparent about data retention policies and allow users to delete or anonymize their data when so desired.

Beyond collection and storage, the use of the data introduces substantial ethical considerations. How this data is analyzed and utilized can have profound implications. For instance, AI systems might use personal information to predict behaviors or suggest interventions. While predictive analytics can tailor more effective and personalized support, it also opens up avenues for potential misuse. The risk of profiling and discrimination exists if the data is employed inappropriately or falls into the wrong hands. Thus, building ethical guidelines into AI development practices is crucial.

Moreover, there is the issue of consent. AI systems must be designed to obtain informed consent from users. Traditional consent mechanisms might not be adequate for everyone, particularly for individuals with cognitive disabilities. Here, adaptive consent processes, which are tailored to the comprehension levels and preferences of the user, become vital. This ensures that all users are fully aware of and comfortable with the ways AI interacts with their personal data.

The regulatory landscape is another critical aspect of the privacy concerns surrounding AI and disability. There are various regulations globally, such as the General Data Protection Regulation (GDPR) in Europe and the Health Insurance Portability and Accountability Act (HIPAA) in the United States, that address aspects of data privacy. However, the rapid advancement of AI technologies often outpaces the development of these regulations, creating gaps that may leave users vulnerable. It's essential for policymakers to stay abreast of technological innovations and continuously update legal frameworks to address new privacy challenges.

In addition to governmental regulations, there's a need for industry standards and best practices. Organizations and developers should adhere to ethical codes and guidelines that prioritize user privacy and security. Incorporating privacy by design principles can help ensure that privacy considerations are integrated into every stage of AI system development— from initial design through to deployment and beyond. Such proactive measures can mitigate potential risks and build greater trust among users.

Transparency also plays a pivotal role in addressing privacy concerns. Users should be provided with clear, accessible information about how AI systems work and the implications of their data use. This includes straightforward and understandable privacy policies and real-time notifications about data collection activities. Empowering users with information enables them to make more informed decisions about their interactions with AI technologies.

Also, consider the role of user feedback and community involvement. By engaging with the disability community—and users of AI technologies specifically—developers can better understand the unique privacy concerns and needs of these individuals. This participatory approach not only helps in creating more effective and respectful AI systems but also fosters a sense of ownership and trust among users.

The development of personalized settings within AI systems is another effective measure. Giving users fine-tuned control over what data is collected and how it is used can significantly alleviate privacy concerns. For example, enabling users to set preferences for data sharing, or to turn off certain data-collection features, provides a customizable and user-centric approach to privacy.

Another significant aspect is the cross-border nature of data flows in the context of AI applications. Often, the data collected by an AI system might be stored or processed in a different country, subjecting it to another set of laws and regulations. This international dimension complicates enforcement and compliance of privacy standards. For AI developers and companies, ensuring that their data practices conform to the highest privacy standards globally is a challenging yet critical requirement.

Lastly, it's worth noting the potential psychological impact of privacy concerns. Constant surveillance and data collection can lead to feelings of insecurity and anxiety among users. For individuals with disabilities who are already navigating a complex array of challenges, these additional stressors can significantly impact their overall well-being. Thus, addressing privacy concerns isn't merely a technical or legal issue; it's also about ensuring the holistic well-being of users.

In conclusion, while AI holds remarkable potential to revolutionize the lives of individuals with disabilities, it's imperative that privacy concerns are rigorously addressed. This requires a multifaceted approach that includes robust data protection measures, strict ethical guidelines, adaptive consent mechanisms, and continuous user engagement. By prioritizing privacy, we can ensure that AI technologies empower users without compromising their dignity and autonomy.

Fair and Inclusive AI Development

As we delve deeper into the intersection of AI and disability, we must address the ethical obligations inherent in developing these technologies. A fair and inclusive AI development process ensures that the benefits of AI are accessible to everyone, particularly those who have historically faced barriers due to disabilities. This section focuses not just on the technical nuances, but on the moral imperatives we must uphold to make AI a truly equitable tool.

Inclusive AI development begins with representation. Historically, people with disabilities have been underrepresented in technological development, which often leads to products and solutions that don't fully address their needs. Ensuring that people with disabilities are involved in every stage of AI development—from ideation to deployment—is crucial. Their firsthand experiences and insights guide the creation of more effective, empathetic, and useful technologies.

Further, inclusive AI must work within a framework of accessibility. This means adhering to established guidelines and standards that govern accessible design, as well as pushing the boundaries to create new, innovative solutions. For instance, AI systems should be designed to work seamlessly with existing assistive technologies like screen readers, voice recognition software, and adaptive hardware. Accessibility should not be an afterthought but a fundamental element of the development process.

Another essential aspect is the elimination of bias in AI systems. AI algorithms learn from data, and if that data contains biases, the AI systems will perpetuate those biases. It is imperative to scrutinize the datasets used to train AI models, ensuring they are diverse and representative of the full spectrum of human experiences, including those of people with disabilities. This involves not only technical solutions like bias detection algorithms but also an ongoing commitment to ethical vigilance.

The principle of fairness extends to the deployment of AI technologies as well. Regulatory frameworks must ensure that these technologies are accessible to all, regardless of socioeconomic status, geography, or other barriers. Policymakers should collaborate with technologists, disability advocates, and other stakeholders to create legislation that mandates inclusive practices and penalizes exclusionary designs.

Partnerships play a crucial role in fostering inclusive AI development. Collaborations between tech companies, academic institutions, non-profit organizations, and governmental bodies can help pool resources, expertise, and perspectives. These partnerships enable the development of AI technologies that are not only innovative but also deeply rooted in the real-world needs of people with disabilities. Joint initiatives can focus on research, pilot projects, and widespread implementation of inclusive AI solutions.

An inclusive approach also requires transparent communication about the capabilities and limitations of AI technologies. Users need to understand how AI systems make decisions, especially in critical areas like healthcare, education, and employment. Transparency builds trust and ensures that users feel confident in adopting these technologies. It also helps to identify and address any shortcomings in the system, fostering continuous improvement.

Education is another cornerstone of fair and inclusive AI development. Raising awareness about the potential of AI to enhance the lives of people with disabilities can inspire new generations of technologists and policymakers to prioritize inclusivity from the outset. Educational programs, workshops, and seminars can provide valuable training on best practices for inclusive design and development, ensuring that future AI systems are built with fairness embedded in their core.

Moreover, the economic model of AI development should be scrutinized to prevent monopolistic practices that could hinder inclusivity. Open-source platforms and community-driven projects can offer alternative pathways for developing and distributing accessible AI technologies. These models democratize the development process, making it easier for smaller organizations and individuals to contribute to and benefit from AI innovations.

The ethical considerations of data privacy and security are also paramount. People with disabilities, like all individuals, have a right to control their personal data. AI systems must be designed with robust privacy protections to ensure that sensitive information is handled with the utmost care. This involves not only compliance with existing regulations but also proactive measures to safeguard user data against potential breaches.

In addition, developers should strive to create AI technologies that promote independent living and self-determination for people with disabilities. This means designing systems that enhance, rather than replace, human capabilities. AI can support individuals in making their own decisions, managing their daily lives, and participating in their communities, fostering a sense of autonomy and empowerment.

Ultimately, the goal of fair and inclusive AI development is to create a world where technology serves as an equalizer, breaking down barriers and opening up new opportunities for everyone. It is a collective responsibility that requires ongoing commitment, collaboration, and innovation. By prioritizing inclusivity, we can harness the transformative power of AI to build a more just and equitable society.

Chapter 15:
Community and Social Inclusion

Integrating AI into community and social frameworks has the power to significantly enhance the quality of life for individuals with disabilities, fostering a sense of belonging and engagement. By leveraging AI, people with disabilities can overcome social isolation, making meaningful connections through tailored platforms that promote active participation in community activities. Accessible virtual spaces can bridge geographical barriers, allowing users to join interest groups, attend events, and engage in social activities they might otherwise miss. Furthermore, AI can aid in organizing community resources and support networks, ensuring that individuals get the assistance they need promptly and efficiently. This synergy of AI and community support not only democratizes access to social opportunities but also empowers individuals to be active, contributing members of society, cultivating an inclusive and supportive environment for all.

AI for Community Engagement

Artificial intelligence (AI) unearths vast potentials for reshaping how communities engage and interact across diverse spaces. By integrating AI technologies into platforms and programs, we can enhance inclusivity and foster dynamic, supportive environments that serve the specific needs of individuals with disabilities.

One impactful way AI assists community engagement is through tailored communication tools. These tools deliver personalized experiences, ensuring seamless interactions regardless of the user's abilities. For instance, AI-driven chatbots can assist in real-time conversations, guiding people through various services and programs within their community. By utilizing natural language processing (NLP), these chatbots can adapt to different communication styles, whether through text, voice, or sign language, making community resources more accessible than ever before.

Furthermore, AI is revolutionizing access to information. Often, local events, social gatherings, and communal activities are under-publicized for people with disabilities due to barriers in traditional communication channels. AI algorithms can sift through vast amounts of data, picking up relevant events and opportunities that are tailored to personal preferences and accessibility needs. These smart recommendation systems ensure that no one is left in the dark about events that could enhance their social involvement and quality of life.

Beyond communication and information access, AI has a powerful role in personalizing community experiences. Imagine a public park where sensors and AI systems work together to create an inclusive environment for people with varying physical abilities. Smart paths equipped with real-time data collection can notify parkgoers about the crowd levels, noise, and other factors that might impact their visit, letting them navigate the park with increased autonomy. Such innovations encourage participation by ensuring that the environment accommodates everyone's needs.

Another significant aspect is the use of AI to foster social connections. Social isolation is a considerable challenge for many with disabilities. AI-driven platforms can match users with similar interests, facilitating social interactions and creating friendships. These systems can go beyond simple matchmaking by organizing virtual meetups and activities, designed with accessibility in mind. Through AI, we can cultivate community bonds that transcend physical and social barriers.

Moreover, it's essential to discuss how AI enhances community support networks. Caregivers, social workers, and community organizers often struggle with the diverse and dynamic needs of their communities. AI tools can analyze real-time data and suggest optimal ways of extending support. For instance, AI can identify members needing immediate assistance or predict future needs, allowing communities to preemptively address issues. This predictive capability can significantly enhance the efficiency of support networks, ensuring timely and relevant assistance.

AI also plays a transformative role in inclusive event planning and attendance. Event organizers can leverage AI to assess their venues for accessibility features such as wheelchair ramps, audio descriptions, and sign language interpreters. AI-driven applications can guide users through these events, providing live translations, captions, and navigation aids. These technologies make events more approachable and enjoyable for everyone, enhancing community spirit and involvement.

In schools, libraries, and other community centers, AI can introduce new learning and recreational methods that cater to all. Interactive AI systems can offer personalized learning experiences, adjusting difficulty levels and topics to match individual learners' needs. Educational AI applications can also provide extracurricular activities tailored to diverse interests and abilities, enriching the overall community experience and fostering inclusive education environments.

When it comes to online communities, AI's role is equally compelling. Digital platforms powered by artificial intelligence can ensure that online spaces are welcoming and accessible. AI algorithms can identify and filter out toxic behavior, making online communities safer and more inclusive. Furthermore, these platforms can offer accessibility features such as speech-to-text and text-to-speech conversions, real-time language translation, and user interface adjustments to accommodate different accessibility needs, ensuring that everyone can participate interactively.

AI's ability to simulate human-like interactions also proves beneficial in creating more inclusive community platforms. Virtual community centers equipped with AI avatars can provide assistance, answer queries, and guide users through various resources. This can be especially useful for those who may find face-to-face interactions daunting or overwhelming. These AI avatars can be programmed to recognize and respond to a range of emotional cues, providing a supportive and understanding interface for all users.

Additionally, AI's predictive analytics capabilities can help communities plan for the future more effectively. By analyzing demographic data, social trends, and usage patterns, AI can forecast the evolving needs of the community. This foresight allows planners and organizers to develop initiatives and programs that are proactive rather than reactive. Such strategic planning can be crucial in ensuring continuous improvement in community engagement and inclusion.

Moreover, AI's capacity to facilitate better emergency response systems is noteworthy. During crises, quick and effective communication is vital. AI can assist in disseminating timely information tailored to the needs of individuals with disabilities. Real-time data analysis during emergencies can also direct resources more efficiently, ensuring that all community members receive the support they need promptly. These advancements can significantly enhance the resilience and robustness of community safety nets.

Furthermore, AI-driven participatory platforms can empower people with disabilities to voice their opinions and contribute to community decision-making processes. AI can facilitate virtual town halls where ideas and concerns are discussed openly, thus including everyone in policy formulation and community planning. Such inclusive approaches ensure that diverse perspectives are considered, leading to more holistic and equitable community outcomes.

Integration of AI in community engagement requires thoughtful implementation with inclusivity at its core. Developers and planners must involve people with disabilities in the design and testing phases to ensure that AI applications genuinely meet their needs. This user-centered approach can lead to technologies that are not only functional but also empowering, providing users with a sense of agency and belonging in their communities.

As we move forward, it is imperative to recognize the importance of ethical considerations in developing AI for community engagement. Transparency, fairness, and accountability should guide the deployment of AI technologies. Communities must be educated about how AI systems work and their potential impacts. Establishing trust in AI technologies is essential for their widespread acceptance and effective use in community settings.

To sum up, AI holds enormous promise for enhancing community engagement for people with disabilities. By breaking down communication barriers, providing personalized experiences, fostering social connections, enhancing support networks, and ensuring event and online inclusivity, AI paves the way for more inclusive and supportive communities. The vision of a community where everyone feels included, valued, and connected is within reach, thanks to the transformative power of artificial intelligence.

Enhancing Social Interaction

In our technology-driven world, enhancing social interaction is crucial for fostering an inclusive society, especially for individuals with disabilities. Social connections can significantly impact mental health, emotional well-being, and overall quality of life. As we've explored various aspects of artificial intelligence (AI) and its practical applications, it's apparent that AI holds immense potential to revolutionize how people with disabilities engage with their communities and the broader social fabric.

For many individuals, social interactions can present a series of hurdles—ranging from physical barriers to communication challenges. AI breaks down these barriers by offering tailored solutions that cater to individual needs. Imagine an AI-powered assistive device that adapts to the unique ways a person communicates and interacts with others. These technologies can ensure that everyone, regardless of their abilities, has the opportunity to form meaningful social connections.

Take, for example, the advancements in natural language processing and conversational AI. These technologies enable the creation of virtual companions and social robots designed to engage in natural, meaningful conversations. These AI-driven companions can help reduce feelings of loneliness and social isolation, providing not just company but also a platform for practice and improvement of social skills. People with speech impairments or social anxiety can benefit immensely from such innovations, as these virtual companions can be programmed to respond with patience and understanding.

Social media platforms are another domain where AI plays a transformative role. Social interaction is increasingly moving online, and AI can ensure that these platforms are accessible to everyone. Voice recognition and transcription technologies convert spoken words into text, making communication easier for individuals with hearing impairments. Similarly, image recognition tools can describe visual content to those with visual impairments, ensuring that they aren't left out of the conversation.

AI isn't just a tool for direct communication; it also enhances social interaction by promoting greater inclusion in community activities. Community events can often be navigational nightmares for people with mobility issues. AI-powered navigation apps equipped with real-time, accessible route planning capabilities can guide attendees through crowded spaces effortlessly. Augmented reality (AR) applications can overlay information about the environment, providing real-time cues for navigating social gatherings.

Moreover, AI-driven applications are now facilitating more accessible public spaces. Smart, AI-integrated environments can adjust to individual needs, enhancing the overall social experience. For instance, adaptive lighting can change based on the specific needs of individuals with sensory processing issues, while smart sound systems can provide clear, localized audio for those with hearing impairments. These small, often imperceptible, tweaks lead to a more inclusive and welcoming environment for everyone.

Another aspect worth highlighting is the role of AI in enabling remote social interactions. With the rise of virtual reality (VR) and augmented reality (AR), immersive virtual social experiences are now more achievable than ever before. Virtual meetups, supported by AI, can offer lifelike social interaction opportunities for people who might find it difficult to leave their homes. These interactions can bolster a sense of community and belonging, significantly impacting the quality of life.

The gamification of social interaction is yet another innovative avenue where AI demonstrates its value. People with social challenges can engage in AI-powered socialization games that mimic real-world interactions, providing a safe space to practice and develop social skills. Gamified social platforms, often employing AI to adapt to user progress, can motivate and reward individuals for engaging more and improving their social competencies.

Additionally, organizations and workplaces can leverage AI tools to create more inclusive social settings. AI can facilitate remote work and collaboration, breaking down geographical barriers and enabling seamless interaction among team members. Virtual team-building activities and social interactions designed by AI algorithms can foster camaraderie and team spirit, even among remote workers. These tools ensure that no team member feels isolated or disconnected due to physical or communication impairments.

Let's not overlook the importance of AI in providing social support networks. AI-powered platforms can connect individuals with disabilities to support groups, both locally and globally. These networks offer resources, share experiences, and provide emotional support, becoming lifelines for many. AI can also match individuals with mentors or buddies who share similar experiences, promoting peer support and enhancing social networks.

AI's role in enhancing social interaction extends to personalized user experiences as well. Machine learning algorithms can adapt social content and interactions to suit individual preferences and needs. For instance, social platforms can use AI to curate feeds that prioritize content relevant to the user's interests and emotional state. This adaptability ensures a more engaging and positive social experience.

While the transformative potential of AI is evident, it's essential to address the ethical considerations and potential risks involved. Ensuring data privacy, preventing misuse of technology, and promoting fair and inclusive AI development are critical. Developers and policymakers must work together to create frameworks that protect users and ensure these technologies are used for the greater good. Ethical AI development that emphasizes transparency and accountability will pave the way for safe and effective use of AI in enhancing social interaction.

Looking at future trends, we can anticipate even more sophisticated AI applications designed to enrich social lives. Emerging technologies like emotion recognition and sentiment analysis will refine how AI perceives and responds to human interactions. These advancements will enable AI systems to offer more empathetic and supportive responses, tailoring interactions to the user's emotional state. For people who struggle with reading social cues or staying attuned to the emotional context of conversations, these technologies could be game-changers.

Furthermore, integrating AI with other emerging technologies like the Internet of Things (IoT) will unlock new possibilities for social interaction. Smart homes and community spaces outfitted with interconnected devices can facilitate social activities tailored to the needs and preferences of their occupants. Imagine a community center that automatically organizes and promotes inclusive social events based on the interests and schedules of its members, all managed by AI.

Lastly, it's worth mentioning the role of continuous feedback and improvement in these technologies. AI systems thrive on data, and user feedback is critical to their evolution. Encouraging users to share their experiences will enable developers to refine and enhance AI applications continually. This dynamic, user-centered approach ensures that AI technologies remain relevant and effective in fostering social inclusion.

In conclusion, AI has the profound capability to enhance social interaction for individuals with disabilities, transforming challenges into opportunities for connection and engagement. By breaking down barriers and creating new avenues for interaction, AI promotes a society where everyone has the chance to participate fully and meaningfully. The future is not just about more advanced technology but about more connected, inclusive communities where social interaction is enriched through the thoughtful application of AI.

Chapter 16:
User-Centered Design

When we talk about creating AI technologies that truly empower people with disabilities, user-centered design isn't just a nice-to-have—it's a necessity. The core principle is understanding and involving the very users these innovations are meant to assist. By prioritizing the needs, preferences, and feedback of people with disabilities right from the initial stages, developers can craft solutions that are not only functional but deeply enriching. It's about more than accessibility; it's about fostering independence, boosting confidence, and improving the quality of life in very tangible ways. This approach ensures that technology doesn't just adapt to users but evolves with them, creating a dynamic, ongoing dialogue that keeps their voices at the forefront of innovation.

principles of inclusive design

Inclusive design isn't just a buzzword; it's a philosophy that centers on designing products, environments, and technologies accessible to as many people as possible, including those with disabilities. At the heart of this approach is user-centered design, which prioritizes the needs, preferences, and experiences of end-users throughout the design and development process. While the concept can be applied to various fields, its significance in AI and assistive technologies for people with disabilities cannot be overstated. By ensuring that inclusivity is a guiding principle, we can unlock the transformative potential of AI to enrich lives and foster independence.

An integral aspect of inclusive design is understanding diversity in user needs. Disabilities manifest in myriad ways—visually, hearing, cognitively, physically, and emotionally. Designing with this diversity in mind requires a nuanced understanding of each group's unique challenges and requirements. This is where user-centered design shines. Unlike traditional design methods that may focus solely on aesthetics or functionality for the average user, user-centered design involves engaging with real users, gathering their feedback, and iterating based on their insights. This process ensures that the end product resonates well with its intended users, offering solutions tailored to their specific needs.

One of the foundational principles of inclusive design is empathy. Designing for people with disabilities demands a deep empathy for their experiences and challenges. This empathy drives the designer to not just understand but to anticipate potential barriers and find innovative solutions. Take, for instance, the development of voice-activated assistants. These devices weren't just created to offer convenience; they were also designed to provide people with mobility impairments a means to interact with technology and manage daily tasks without needing physical input. The very essence of these technologies lies in seeing the world from the user's perspective.

Flexibility and adaptability form another critical pillar of inclusive design. Solutions must cater to a wide range of abilities and preferences. This adaptability can take many forms—configurable interface settings, customizable hardware options, or scalable features that grow with the user's needs. In practice, this might mean designing software that offers multiple input methods (voice, touch, gesture) or a website optimized for screen readers and keyboard navigation for visually impaired users. The goal is to create adaptable environments that can quickly respond to varying needs, ensuring no one is left behind.

Inclusivity also emphasizes simplicity and user-friendliness. Complex interfaces or convoluted processes can be overwhelming, especially for users with cognitive disabilities. Simplified navigation, clear instructions, and intuitive design are essential for creating accessible technology. User-centered design principles advocate for minimizing the cognitive load on users, making sure that interactions are straightforward and require minimal effort. These considerations are crucial not only for usability but also for building confidence and fostering independent use among people with disabilities.

Communication and collaboration form the bedrock of effective user-centered, inclusive design. Involving end-users in the prototyping and testing phases ensures that the product development cycle remains responsive to their needs. Regular feedback loops and real-world testing allow designers to identify and rectify potential issues before they become major barriers. This collaborative approach cultivates a sense of ownership among users and empowers them to be actively involved in shaping the technologies they rely on.

Moreover, ethical considerations must underpin every stage of the inclusive design process. Issues like privacy, data security, and user autonomy are paramount, particularly for AI technologies that process sensitive personal information. An ethical framework ensures that technologies not only serve but also protect their users. Transparency, consent, and fairness are ethical cornerstones that must be integrated into the design. For example, AI algorithms should be transparent about how they make decisions, and users should have control over their data and how it is used.

Another key principle is sustainability. Sustainable design is about creating durable, reliable products that don't just serve immediate needs but are built to adapt and evolve. This foresight is vital for users with disabilities, whose needs might change over time. Sustainable design models account for future adaptability, ensuring that initial investments in technology continue to yield benefits in the long run. This might entail building modular designs that can be easily upgraded or creating software that's compatible with future updates and technological advancements.

Education and awareness are also vital components. Creating inclusive AI technologies isn't just the responsibility of developers and designers; it involves educating all stakeholders, including policymakers, end-users, and even the general public. Awareness initiatives can demystify AI and assistive technologies, shedding light on their benefits and potential pitfalls. Educational efforts also pave the way for a more inclusive societal mindset, encouraging wider adoption and support for these technologies.

Interdisciplinary collaboration is crucial in the realm of inclusive design. Bringing together experts from various fields—engineers, designers, therapists, educators, and users—creates a rich tapestry of insights and perspectives. This collaborative energy can spur innovative solutions that a single-discipline approach might overlook. When cross-functional teams work together, the end products are more holistic, addressing a broader spectrum of needs.

Ultimately, the principles of inclusive design aim to break down barriers and create equitable experiences. By embracing user-centered design, we ensure that technologies don't merely serve the majority but uplift those who might otherwise be marginalized. AI has the power to be a great equalizer, offering opportunities for independence and participation that were previously unimaginable for many people with disabilities. Through dedicated efforts to include all users in the design process, we can create a future where technology truly works for everyone.

Involving Users in Development

Developing AI technologies that empower individuals with disabilities requires more than just technical proficiency. It necessitates a deep and sincere engagement with the very people these tools aim to assist. User-centered design isn't merely a box to check; it's a crucial aspect that dictates the effectiveness of any AI-driven solution. By involving users in the development process, we can create tools that are truly customized to meet the unique and varied needs of people with disabilities.

Engaging users in development starts with understanding their daily challenges and aspirations. This step goes beyond reading research papers and articles. It involves immersive, first-hand experiences. Developers need to spend time with users in their environments, observing their routines, and listening to their stories. Only through this intimate understanding can we pinpoint the nuanced problems that AI solutions should address.

Including users from the outset ensures that the technology is relevant and usable. Initial ideas for AI applications must pass the scrutiny of those who know best—the end users. Their feedback can be invaluable in shaping the direction of the project. For instance, a voice-activated assistant for individuals with speech impairments must be tested for its responsiveness to various speech patterns right from the prototype stage. Ignoring this could lead to a tool that's technically impressive but operationally redundant.

The involvement of users shouldn't end after the initial consultation. Their role is vital throughout the iterative design process. Developing successful AI solutions is inherently an adaptive exercise. User feedback should shape each iteration, ensuring continuous alignment with real-world needs. Regular feedback loops enable swift identification and rectification of any issues, preventing costly rework.

It's also important to create diverse user groups. People with disabilities are not a monolith. They have a range of experiences, backgrounds, and needs. A visually impaired user from an urban area will have different challenges compared to one from a rural setting. Ensuring diversity in user testing groups will help develop more versatile and adaptable AI solutions.

Leveraging user insights also stimulates innovation. Often, users will suggest features or improvements that may never occur to developers. Take AI-driven communication aids, for instance. Users might highlight the importance of integrating social media capabilities, something the developers hadn't previously considered. By tapping into the collective creativity of users, we can uncover groundbreaking features that elevate these technologies to new heights.

Understanding user preferences extends to the interface and usability aspects of AI applications. Sometimes, the success of an AI tool hinges on how intuitively it can be used. Users can provide precise feedback on the ease of navigating through the application, the readability of fonts, or the accessibility of buttons and controls. Simple yet vital adjustments based on this input can significantly enhance user experience.

One critical aspect of involving users is respect and empathy. When engaging with individuals with disabilities, it's vital to approach with humility, recognizing that they are the experts of their own experiences. Building trust is fundamental. Users need to feel that their perspectives are valued and will genuinely influence the outcome. This can be achieved by transparent communication and by showing tangible results from their contributions.

Empowering users as co-creators in the development process not only leads to more effective solutions but also fosters a sense of ownership and advocacy. When users see their input materialized in the final product, it strengthens their commitment to the tool's success. They become advocates and ambassadors for the technology, helping to promote it within their networks.

Involving users doesn't have to be a resource-intensive exercise. Various methods exist for engaging users efficiently. Workshops, focus groups, and online forums can serve as platforms for gathering user insights. Digital tools, such as surveys and usability testing software, can reach a wider audience and collect diverse feedback swiftly. Even remote usability sessions can be highly effective, providing flexibility for users who might face transportation challenges.

Another compelling approach is the participatory design, where users are not just testers but active participants in the design team. They collaborate closely with developers, contributing to brainstorming sessions, wireframing, prototyping, and testing. This level of involvement ensures that their needs and preferences are integrated seamlessly into the design.

For developers, this paradigm shift towards user-centered design requires a change in mindset. They need to evolve from the traditional top-down approach where technology is imposed on users, to a bottom-up method where user insights drive the technology. This cultural shift within development teams can be facilitated by training programs and workshops focused on empathy, active listening, and inclusive design practices.

Educational institutions can play a pivotal role by embedding user-centered design in their curriculum. Future engineers, designers, and technologists should be groomed to think inclusively from the outset. Courses on inclusive design, accessibility standards, and ethics in AI should become cornerstones of tech education. By instilling these values early, we can build a generation of developers who are inherently aligned with the principles of inclusive and user-centered design.

The role of user-centered design doesn't end with the deployment of AI technologies. Continuous user engagement is critical for ongoing improvements and updates. Users should have avenues to report issues, suggest enhancements, and share their evolving needs. Long-term relationships between developers and user communities can ensure that the technology remains relevant and effective over time.

To summarize, involving users in the development process is not a mere formality but a fundamental ingredient for the success of AI technologies aimed at empowering individuals with disabilities. A nuanced understanding of user needs, continuous engagement through iterative design processes, respect, and empathy towards users, and fostering a culture of inclusivity within development teams are all crucial steps. When done right, this collaborative approach can result in AI solutions that truly enhance accessibility, independence, and quality of life for people with disabilities, transforming visions into reality.

Chapter 17: Challenges and Barriers in Implementation

Despite the advancements in artificial intelligence geared towards empowering people with disabilities, the path to effective implementation is fraught with challenges. Technical barriers such as the lack of diverse, high-quality training data, and the complexities of developing adaptive algorithms that cater to various disabilities are significant hurdles. Additionally, societal barriers like limited awareness, reluctance to adopt new technologies, and financial constraints often impede widespread accessibility. Ensuring that AI systems are designed inclusively from the ground up is crucial but requires collaboration among developers, stakeholders, and the disabled community. By understanding these challenges and actively working to overcome them, we can make AI applications not just innovative, but universally beneficial.

Technical and Societal Challenges

Implementing AI technologies to enhance the lives of people with disabilities isn't a straightforward task. Both technical and societal challenges can act as substantial barriers. Understanding these obstacles is crucial to overcoming them and making meaningful progress.

On the technical side, one of the biggest challenges is data. AI systems rely heavily on vast amounts of data to learn and make predictions. For disabilities that are less common, gathering sufficient data is a hurdle. Even when data is available, it often lacks diversity. This makes it difficult for AI models to generalize across different populations and conditions. Moreover, labeled data, which is essential for supervised learning, is particularly scarce and expensive to obtain.

Hardware is another issue. AI-powered assistive technologies often need specialized hardware to function optimally. Developing custom hardware that meets the unique needs of people with disabilities can be costly and time-consuming. This issue is compounded by the rapid pace of technological advancements. Continuously updating hardware to keep pace with innovations in AI can be a logistical nightmare.

Beyond data and hardware, the complexity of human problems poses another significant challenge. Disabilities are incredibly varied and personal. Building AI systems that can adapt to this variability is an enormous task. For example, two individuals with the same type of disability might have entirely different needs and preferences. Creating one-size-fits-all solutions simply doesn't work in this context. The AI has to be highly customizable, which adds layers of complexity to the development process.

Algorithmic bias is a persistent issue. AI systems are only as good as the data they are trained on. If that data includes biases—whether explicit or implicit—the AI will perpetuate those biases. For people with disabilities, this can result in technologies that are not only ineffective but potentially harmful. Ensuring that AI systems are fair and unbiased requires rigorous testing and constant vigilance.

Societal challenges are just as daunting. One major issue is the lack of awareness and understanding. Many people, including those who design and implement AI systems, don't fully understand the challenges faced by people with disabilities. This results in solutions that are well-intentioned but ultimately ineffective.

Stigmatization is another significant barrier. People with disabilities often face societal stigma, which can hinder the adoption of new technologies. If an AI solution highlights the user's disability in a way that feels intrusive or stigmatizing, it will likely be rejected. Any effective implementation needs to be sensitive to these social dynamics.

Economic factors also play a role. The cost of developing and deploying AI-powered assistive technologies can be prohibitive. Many individuals with disabilities have limited financial resources, making it difficult for them to afford such technologies. Even when the technology is available, it may not be accessible due to cost barriers. Public and private sector funding can help alleviate this, but financial limitations remain a significant obstacle.

Legislative and regulatory environments can also be slow to adapt. Laws and regulations often lag behind technological advancements. For AI technologies, this means navigating a complex and outdated regulatory landscape. In some cases, existing laws might even hinder the development and deployment of new technologies. Advocating for updated regulations that consider the unique needs of people with disabilities is essential for progress.

Ethical considerations can't be ignored. As AI increasingly becomes a part of daily living, ethical questions around privacy, consent, and autonomy come to the fore. It's crucial to ensure that these technologies empower people with disabilities rather than control them. Building ethical AI involves making difficult trade-offs and constantly seeking input from the communities affected.

A successful implementation of AI technologies requires multidisciplinary collaboration. It's not enough for engineers and data scientists to work in isolation. Input and expertise from healthcare professionals, educators, and individuals with disabilities are critical. This collaborative approach can help ensure that solutions are both effective and culturally sensitive.

Involving users in the development process is another key strategy. People with disabilities should not be passive recipients of technology but active participants in its creation. User-centered design principles can help make sure that the end product actually meets the needs of its intended users. This involves iterative testing and feedback, which can be time-consuming but ultimately leads to better outcomes.

Public perception and trust in AI also play a significant role. There's often skepticism around new technologies, and AI is no exception. Building trust requires transparency and education. People need to understand how these technologies work, what their benefits are, and how their data will be used. Open communication can help demystify AI and build public confidence.

Sustainability is another critical factor. Once an AI system is deployed, it needs to be maintained and updated. This requires ongoing resources, both financial and human. Sustainability plans should be part of the initial design process, ensuring that the technology remains effective over the long term.

While the challenges are numerous, they are not insurmountable. Addressing both technical and societal barriers is essential for the successful implementation of AI technologies that can truly enhance the lives of people with disabilities. A proactive, inclusive, and collaborative approach can help overcome these obstacles and pave the way for a more accessible and equitable future.

Addressing Accessibility in AI Systems

When it comes to implementing AI systems, especially for enhancing accessibility, several formidable challenges emerge. Despite the potential benefits, the path isn't always straightforward. Accessibility in AI is not merely about adding features; it's about rethinking entire systems to ensure inclusivity from the ground up. Whereas society has begun to recognize the significance of accessible environments in the physical world, implementing these ideals into the digital and AI spaces remains a continuously evolving challenge.

One prominent barrier is the technical complexity of developing AI systems that can cater to diverse needs. Disabilities are not monolithic; they range widely in type and severity. An AI system that works well for one group might not be suitable for another. Creating algorithms that are adaptable and responsive to such a broad spectrum necessitates advanced machine learning techniques and substantial data sets. Collecting this data responsibly while safeguarding privacy poses yet another layer of complexity.

Data diversity is a critical issue here. AI systems trained predominantly on data from non-disabled users may inadvertently exclude or misrepresent individuals with disabilities. This is a common pitfall in current AI systems where biased data leads to biased outcomes, reinforcing existing disparities instead of narrowing them. To combat this, it's essential to actively include data from people with disabilities during the training phases, even though this can involve logistical and ethical challenges.

Moreover, the classical notion of "one-size-fits-all" in software development couldn't be less applicable when it comes to accessibility-oriented AI systems. These systems must offer customization and flexibility to accommodate unique individual needs. For instance, speech recognition technology must understand various accents and speech impediments. Vision-impaired users might need completely different interfaces compared to those with hearing impairments. Tailoring solutions in this manner often multiplies development time and resources, posing significant challenges, especially for small developers.

An often-overlooked barrier is societal attitudes towards disability and technology. Misconceptions about what AI can and can't do can result in a lack of buy-in from potential users and stakeholders. For instance, some may be skeptical of AI's efficacy or worry about potential job losses due to automation. Bridging this gap involves education and advocacy, making users and decision-makers aware of AI's true potential while dispelling myths and uncertainties.

The legal and ethical landscape further complicates things. Regulations around data privacy, bias, and the ethical use of AI aren't always clear-cut, especially in an international context. Developers must navigate a labyrinth of laws and guidelines, which can differ significantly between regions. Ensuring compliance while maintaining the customizability and functionality of AI systems adds another layer of complexity to the development process.

Interoperability is another hurdle. AI systems don't operate in isolation. Integration with existing technologies, platforms, and even daily routines is paramount for seamless user experiences. However, achieving interoperability is easier said than done. Different systems and devices often use varying standards and protocols, making seamless integration a technical challenge. Consequently, both backward compatibility and forward adaptability become key aspects requiring meticulous planning and execution.

User engagement remains a pivotal element in the development and implementation of accessible AI systems. Involving users with disabilities in the design, testing, and feedback loops is essential. However, accessing such user groups and securing their sustained participation involves logistical hurdles and resource allocation. It's not merely about consulting a few users; it's about building long-term relationships that allow for iterative improvement.

Cost is an unavoidable challenge. Developing AI systems, especially those tailored for accessibility, requires significant investment. Often, non-profit organizations and smaller companies may find it financially unfeasible to invest in such development costs without external funding or grants. Even larger corporations must justify these expenditures, frequently sidelining accessibility projects when immediate financial returns seem uncertain.

Add to this the rapid pace of technological advancement. While new developments in AI present opportunities for enhanced accessibility, they also create a moving target for developers. Keeping up-to-date with the latest technologies while ensuring the robustness and reliability of existing systems presents an ongoing balancing act. Moreover, older hardware and software may not support the latest AI innovations, creating a technological divide that can exclude individuals reliant on such equipment.

Accessibility isn't just about technical solutions; it's deeply human-centric. Linguistic barriers, cultural differences, and individual user characteristics all influence how AI interactions unfold. Thus, systems must be adjustable not just in terms of hardware and software but also in cultural and contextual adaptability. For example, idiomatic phrases or cultural references within AI-driven tools must resonate appropriately with diverse user groups, demanding localized content and UX strategies.

Beyond development, the effective deployment of AI systems involves thorough training, support structures, and constant updates to maintain efficacy and security. Both end-users and administrators require training to maximize the benefits of accessible AI tools. Adequate support documentation, real-time assistance, and regular updates form the ecosystem that ensures these systems don't just exist but thrive in real-world applications.

The road to making AI genuinely accessible is strewn with challenges, but it's one worth traveling. The vision of an inclusive world, where technology empowers all individuals regardless of their abilities, is not just aspirational but achievable. By adopting a multi-faceted approach addressing both technical and societal barriers, while fostering a culture of inclusivity and empathy, we can bridge the gaps that currently exist.

Ultimately, AI systems that prioritize accessibility are emblematic of broader values—equality, inclusion, and human dignity. They symbolize our collective effort to harness technology for the greater good, ensuring that innovation benefits everyone, especially those who have been historically marginalized. Addressing accessibility in AI systems isn't merely a technical endeavor; it's a call to action for a more equitable future.

Chapter 18:
Future Trends and Innovations

The future of AI in enhancing accessibility for individuals with disabilities holds unparalleled promise and transformative potential. Emerging AI technologies, from advanced neural networks to intuitive natural language processing systems, are pushing the boundaries of what's possible. These innovations promise to create more inclusive environments by continuously learning and adapting to individual needs. Predicted developments point to AI evolving into ever-more personalized and anticipatory tools, seamlessly integrating with daily life to improve both independence and quality of life. As AI technologies develop, the emphasis on user-centered design and ethical considerations will be paramount to ensure that these advancements are equitable and widely beneficial. This forward-looking perspective not only excites but also inspires a commitment to leveraging AI for a truly inclusive future.

Emerging AI Technologies

In recent years, rapid advancements in artificial intelligence have opened doors to technologies that were once confined to the realm of science fiction. AI is steadily weaving into the fabric of our daily lives and, for individuals with disabilities, it holds transformative potential. Emerging AI technologies are not just about creating high-tech gadgets; they're about fundamentally redefining how people with disabilities interact with the world, providing unprecedented levels of accessibility, independence, and quality of life.

One of the more promising developments is AI-augmented reality (AR). Integrating AI with AR can dramatically enhance sensory experiences. For instance, visually impaired individuals can use AR glasses equipped with AI algorithms to describe their surroundings in real time, identify objects, read text, and even recognize faces. This bridging of visual gaps enables greater independence in daily activities, such as shopping, navigating public spaces, or social interactions.

AI-driven wearable technology is another frontier with transformative capability. Wearables equipped with advanced sensors and AI-powered analytics can monitor vital statistics, detect anomalies, and provide real-time feedback. For people with physical disabilities, exoskeletons enhanced with machine learning algorithms can learn from user movements, adapting to provide smoother, more natural motion support. These intelligent systems can empower users, improving mobility and reducing reliance on caregivers.

Natural language processing (NLP) continues to evolve, bringing forth innovations that can significantly benefit those with speech and language impairments. AI applications can now convert text to speech and vice versa with remarkable accuracy and speed. More so, they provide contextual understanding, enhancing communication tools to be more intuitive and responsive. Conversation AI, capable of understanding nuances and context, can revolutionize how individuals interact with digital assistants and other software, making communication more seamless and effective.

When it comes to cognitive disabilities, AI is paving the way for several groundbreaking tools. AI-based learning platforms can provide personalized learning experiences tailored to each individual's needs, helping those with cognitive challenges process information at their own pace. These platforms use adaptive learning algorithms to continuously adjust the complexity and mode of information delivery based on user interaction, promoting better comprehension and retention.

Another remarkable application of AI lies in predictive analytics for mental health. Machine learning models can analyze patterns in data from various sources, including social media, wearable devices, and health records, to predict mental health episodes. Early identification of such episodes can prompt timely intervention, providing an opportunity to mitigate issues before they escalate. These technologies offer a proactive approach to mental health, giving individuals and caregivers advanced tools for management and support.

AI also plays a critical role in creating immersive and interactive environments for cognitive rehabilitation. Virtual reality (VR) combined with AI can offer simulated environments where individuals can practice daily tasks, social interactions, and even job-related skills. These controlled, safe environments can significantly improve one's confidence and competence before transferring these skills to real-world scenarios.

In the realm of mobility, AI's role in autonomous vehicles stands out. Self-driving cars equipped with sophisticated AI systems can offer reliable transportation options for people with mobility impairments. Such vehicles can navigate complex traffic systems, recognize pedestrians, and adapt to various driving conditions, providing a level of independence previously unimaginable for individuals who cannot drive.

Smart home technologies are also benefiting from AI innovations, further enhancing the independence of people with disabilities. Smart sensors, powered by AI, can recognize user patterns, adjust home settings automatically, and even predict needs. For example, intelligent door locks can identify users, voice-controlled lights and thermostats cater to the user's preferences, and AI-powered security systems can provide an additional layer of safety. These technologies foster a supportive environment that empowers individuals to manage their daily lives more effectively.

Furthermore, AI is revolutionizing the accessibility of digital content. Advanced image recognition algorithms can describe the content of images and videos, making social media and other online platforms more inclusive. For individuals with hearing impairments, AI can provide real-time captioning or even sign language translations, enhancing their ability to consume multimedia content.

As we continue to push the frontiers of AI, ethical considerations must remain at the forefront. These technologies should be developed inclusively, with input from the disabled community to ensure they address real needs and do not introduce new barriers. Accessibility and usability should be baked into the design process rather than considered at the end. Moreover, there needs to be a focus on ensuring that AI systems used in healthcare or as decision aids are transparent, fair, and free from biases that could disproportionately affect people with disabilities.

With these innovative technologies on the horizon, it's clear that AI holds a myriad of possibilities for enhancing the lives of individuals with disabilities. As developers, healthcare professionals, educators, and tech enthusiasts collaborate, it's crucial to remain engaged with the communities we aim to support, ensuring that these advancements translate into meaningful improvements in accessibility and independence.

The future of AI in this realm isn't just about creating new technologies; it's about creating opportunities—opportunities for individuals to lead dignified, independent, and fulfilling lives. The possibilities are exciting and endless, but they require our collective effort to realize their full potential. Emerging AI technologies, when thoughtfully and inclusively applied, can truly be the catalyst for a more accessible and inclusive world.

Predicted Developments

Artificial intelligence is advancing at an unprecedented pace, and its potential to transform the lives of individuals with disabilities is both exciting and profound. As we look towards the horizon, several predicted developments stand to redefine the landscape of assistive technologies. One of the key areas of growth lies in the integration of AI with internet-of-things (IoT) devices. This synergy could lead to enhanced personal environments, where homes and public spaces automatically adjust to the needs of individuals based on real-time data. Imagine a home that can learn and adapt to the accessibility needs of its occupants, providing a tailored experience that significantly boosts independence and quality of life.

Adaptive learning systems powered by AI are expected to become more sophisticated, offering personalized educational journeys. For students with cognitive disabilities, these systems could provide a highly individualized curriculum, recognizing and adapting to unique learning styles and needs. By leveraging massive datasets, AI can pinpoint the specific areas where a student might struggle and offer targeted support, thus bridging educational gaps more effectively.

AI-driven prosthetics and mobility aids are also on the cusp of revolutionary improvements. Current advancements in AI and machine learning are paving the way for prosthetics that can adapt and respond in real-time to the user's movements. This adaptability can provide a more natural and intuitive experience, closely mimicking the user's intent. The integration of AI with robotics is set to enhance the functionality of these devices, offering greater dexterity, strength, and endurance.

Voice recognition and natural language processing (NLP) technologies have already made significant strides, and their future looks even brighter. These technologies will likely become more seamless and intuitive, providing better support for communication aids. For individuals with speech impairments, improved AI algorithms could offer more accurate translations and interpretations, facilitating clearer and more effective interaction.

Emotion-sensing AI is making waves in mental health support, and future developments in this field are poised to offer more nuanced and empathetic assistance. These systems can analyze facial expressions, voice tones, and even physiological signals to gauge emotional states, providing timely interventions or suggesting coping strategies. As emotional AI becomes more sophisticated, it could play a critical role in offering personalized mental health support, potentially preventing crises before they occur.

Healthcare is another domain where AI's impact will grow significantly. Predictive analytics and personalized healthcare models are anticipated to become standard practice. AI can analyze vast amounts of data to predict health trends and outcomes, allowing for early intervention and customized treatment plans. This is particularly crucial for individuals with disabilities who may have complex medical needs that require ongoing management.

AI's role in enhancing workplace accessibility is also forecasted to expand. Technologies that facilitate remote work and collaboration, such as virtual meeting platforms with integrated speech-to-text and language translation features, will continue to evolve. These tools can make workplaces more inclusive, ensuring that individuals with disabilities have equal opportunities to contribute and thrive professionally.

In the realm of mobility, self-driving technology holds substantial promise. Autonomous vehicles could redefine public and private transportation for individuals with mobility impairments. These vehicles, equipped with advanced AI systems, could offer door-to-door transportation, significantly reducing reliance on traditional, less accessible forms of transport.

We can't overlook the potential of AI in fostering community and social inclusion. Innovations in augmented reality (AR) and virtual reality (VR), powered by AI, are likely to create more immersive and inclusive environments. These experiences can help individuals with disabilities to participate more fully in social activities, providing a sense of community and belonging.

Legal and ethical frameworks surrounding AI will evolve alongside technological advancements. It is anticipated that more robust and inclusive policies will emerge, aimed at ensuring that AI developments are fair, transparent, and beneficial for all. This will involve addressing concerns around privacy, data security, and algorithmic bias, ensuring that AI systems are developed and deployed with equity at their core.

The involvement of users in the design and development process of AI technologies is expected to become more commonplace. Co-design approaches will ensure that the technologies are truly user-centered, addressing the real needs and challenges faced by individuals with disabilities. This collaborative approach can lead to more innovative and effective solutions, empowering users to take an active role in shaping the future of assistive technologies.

Global collaboration in AI research and development will be pivotal. As nations share insights and innovations, the cumulative knowledge can propel the development of AI in new and unexpected directions. International partnerships can help standardize best practices and accelerate the deployment of effective AI solutions worldwide.

Finally, the role of AI in personalized learning environments will continue to grow, providing educators with powerful tools to create adaptive and responsive educational experiences. AI can help identify learning patterns, predict outcomes, and tailor educational content to meet the unique needs of each student. This level of personalization holds the promise of making education more accessible and effective for students with disabilities.

In conclusion, the predicted developments in AI for individuals with disabilities are nothing short of revolutionary. The convergence of AI with other advanced technologies, coupled with an inclusive and user-centered approach, holds the potential to dramatically enhance accessibility, independence, and quality of life. As we forge ahead, it's essential to remain focused on creating equitable, ethical, and effective solutions that genuinely empower individuals with disabilities to lead fuller, more autonomous lives.

Chapter 19:
Global Perspectives

In our exploration of how AI technologies can empower people with disabilities, it's essential to recognize the diverse ways these innovations are being adopted and implemented across the globe. Different countries and cultures bring unique perspectives and solutions to the table, enriching the collective understanding and development of AI for accessibility. While some nations innovate with cutting-edge technology and robust support systems, others blend traditional practices with modern advancements to create localized solutions tailored to their communities' needs. From Sweden's extensive use of AI in public accessibility policies to India's grassroots movements integrating AI in education for the visually impaired, these global perspectives highlight the importance of collaboration and knowledge sharing. Learning from these varied approaches not only broadens our horizon but also inspires more effective, inclusive, and compassionate advancements in AI to assist people with disabilities universally.

Case Studies from Around the World

Artificial intelligence (AI) solutions for people with disabilities are being developed and implemented across the globe, enhancing accessibility, independence, and quality of life. By looking at specific examples from diverse nations, we can gain a deeper understanding of how these technologies are tailored to meet local needs and challenges.

These case studies also reveal the cultural, economic, and social factors influencing the development and adoption of AI technologies.

In Japan, a country known for its rapid technological advancement, AI-driven solutions have become integral to assisting the elderly and people with disabilities. One notable example is the use of robotic caregivers. These AI-powered robots, such as the "Robear," are designed to help elderly and disabled individuals with mobility tasks, including transferring from a bed to a wheelchair. Robear combines gentle strength with a friendly appearance, making it less intimidating for users. By significantly reducing the physical strain on caregivers, this innovative technology addresses the shortage of healthcare professionals in an aging society.

In contrast, India has taken a community-driven approach to AI innovations for people with disabilities. Organizations like Eye-D are leveraging AI to develop apps that assist visually impaired individuals in navigating their surroundings. Eye-D, for instance, uses AI to describe the environment, read text, and identify objects. This tool is gaining traction in urban areas where public infrastructure for the visually impaired is often lacking. By providing real-time information about the user's surroundings, Eye-D empowers visually impaired individuals, promoting greater independence and safety.

Sweden, recognized for its inclusive policies, is another leading nation in the application of AI for disability support. The country has been focusing on AI-powered communication devices to support individuals with speech and language impairments. Tobii Dynavox, a Swedish company, has developed eye-tracking technology that allows people with severe disabilities to communicate through text-to-speech applications. This technology has proven to be life-changing for individuals with conditions like ALS, offering them a voice and a means to interact with the world.

In Singapore, the government is actively promoting AI applications to improve the lives of its citizens with disabilities. One standout initiative is the integration of AI in public transport systems. The "Smart Bus Stop" project uses AI to assist visually impaired passengers in identifying bus arrival times and ensuring they board the correct bus. This technology is part of broader efforts to transform Singapore into a "Smart Nation," where digital technology and AI play crucial roles in enhancing accessibility and inclusivity.

Australia has also been at the forefront of utilizing AI to aid individuals with hearing impairments. The country's National Relay Service (NRS) leverages AI to enhance real-time communication between people who are deaf or hard of hearing and those who can hear. Through AI-driven transcription services, NRS offers a seamless conversation experience by converting spoken words into text and vice versa. This service is particularly important in remote areas where access to traditional communication aids may be limited.

In Brazil, researchers are exploring the potential of AI to assist students with cognitive disabilities. The Universidade Federal Fluminense has been developing AI-based educational tools tailored to the needs of children with autism. These tools use machine learning algorithms to adapt educational content according to the child's learning pace and style. By creating a personalized learning environment, these AI applications ensure that students with cognitive disabilities receive the support they need to succeed academically.

The United Kingdom is another leader in integrating AI into accessibility solutions. The UK's National Health Service (NHS) has been investing in AI technologies to support mental health. One pioneering initiative is the use of AI chatbots for mental health monitoring and support. These chatbots can engage in conversations with users, offering immediate emotional support and, if necessary, directing them to professional help. By reducing the stigma associated with mental health issues and providing accessible support, these AI-driven solutions are making a significant impact.

Canada is leveraging AI to promote inclusive employment for people with disabilities. The Canadian organization, Magnet, uses AI to match job seekers with disabilities to suitable job opportunities and employers who prioritize inclusive hiring practices. The platform analyzes the skills, experiences, and preferences of users and then uses machine learning algorithms to provide personalized job recommendations. This targeted approach helps bridge the employment gap for people with disabilities, promoting economic independence and social inclusion.

In Kenya, AI technologies are being developed to address unique challenges faced by people with disabilities in low-resource settings. For instance, the iHub research lab has developed AI-powered sign language interpreters to assist the deaf community. This tool uses computer vision to translate sign language into text, making communication more accessible for those who cannot sign. By localizing technology to meet the specific needs of the community, such initiatives are making significant strides in improving the quality of life for people with disabilities in Kenya.

Germany is known for its robust healthcare system and technological innovation. The country has been incorporating AI into rehabilitation processes for individuals with physical disabilities. One example is the use of AI-driven exoskeletons, which help patients regain mobility. These devices use machine learning algorithms to adapt to the patient's movements, providing customized support and improving rehabilitation outcomes. By enabling individuals to regain independence and participate more fully in society, AI-driven rehabilitation technologies are transforming lives in Germany.

In South Korea, the focus has been on developing AI applications for inclusive education. The Korean Institute for Special Education has created AI-based learning platforms that cater to students with diverse learning needs. These platforms offer personalized learning experiences by using AI to adapt lessons based on individual progress and comprehension levels. By ensuring that all students, regardless of their abilities, have access to quality education, South Korea is promoting a more inclusive society.

In Israel, AI is being used to develop innovative assistive technologies for individuals with mobility impairments. ReWalk Robotics, an Israeli company, has created an exoskeleton that helps paraplegics stand and walk. This AI-powered device uses sophisticated sensors and control algorithms to mimic natural walking patterns, offering users greater mobility and independence. By leveraging AI to enhance the functionality and usability of assistive technologies, Israel is making significant contributions to the field.

China is also making strides in using AI to support people with disabilities. One notable project is the development of smart glasses that assist visually impaired individuals. These glasses use computer vision and machine learning to recognize objects, read text, and provide navigation assistance. By offering a wearable and intuitive solution, these smart glasses are empowering visually impaired individuals to navigate their environment more confidently and independently.

These global case studies highlight the diverse ways in which AI technologies are being applied to enhance the lives of people with disabilities. They underscore the importance of tailoring solutions to meet local needs, reflecting cultural, social, and economic contexts. From robotic caregivers in Japan to personalized learning platforms in South Korea, these examples demonstrate the transformative potential of AI in promoting accessibility, independence, and quality of life for people with disabilities worldwide.

The case studies also illustrate the collaborative efforts between governments, organizations, and the private sector in driving these innovations. By sharing knowledge and best practices, countries can learn from each other and accelerate the development and adoption of AI solutions. This global perspective not only fosters a more inclusive digital landscape but also ensures that the benefits of AI reach all corners of the world.

International Standards and Practices

In the realm of artificial intelligence (AI) for people with disabilities, international standards and practices serve as the backbone for creating consistent, reliable, and equitable solutions worldwide. These frameworks ensure that AI technologies are developed and implemented in ways that respect human rights, prioritize accessibility, and foster inclusivity. Different countries and organizations have enacted a variety of standards aimed at promoting these principles, creating a patchwork of approaches that reflect diverse cultural, legal, and social contexts.

One cornerstone in establishing international standards for AI is the involvement of global organizations such as the International Organization for Standardization (ISO) and the World Wide Web Consortium (W3C). The ISO, with its vast array of technical standards, has been pivotal in ensuring that assistive technologies meet rigorous quality and safety benchmarks. Standards such as ISO 9241, which focuses on ergonomic requirements for office work with visual display terminals, indirectly impact accessibility features in AI-driven tools. The W3C, on the other hand, has championed internet accessibility through its Web Content Accessibility Guidelines (WCAG), which are essential for AI applications that involve web interfaces.

From a policy perspective, the United Nations (UN) Convention on the Rights of Persons with Disabilities (CRPD) has been a significant driver in aligning international efforts toward inclusive AI. Articles in the CRPD emphasize the importance of accessibility and the role of assistive technologies in enabling people with disabilities to live independently and participate fully in all aspects of life. This convention serves as a critical reference for national legislations and regulatory practices aimed at promoting inclusive AI.

Individual countries have also taken steps to develop and enforce their own accessibility standards and regulatory frameworks. In the United States, the Americans with Disabilities Act (ADA) and Section 508 of the Rehabilitation Act require that digital and electronic information technology be accessible to people with disabilities. These laws have a significant influence on how AI technologies are designed and implemented, often requiring compliance to avoid legal repercussions.

In the European Union, the European Accessibility Act (EAA) and the General Data Protection Regulation (GDPR) shape the landscape of AI accessibility and data privacy. The EAA mandates that certain products and services, including AI technologies, must be accessible to people with disabilities, while the GDPR ensures that personal data is handled with rigorous privacy protections. Together, these regulations create a robust framework that influences AI development and deployment across the continent.

Asia-Pacific countries are also making strides in establishing accessibility standards for AI. Japan, for example, has adopted the Japanese Industrial Standards (JIS) based largely on ISO standards, focusing heavily on workplace and educational accessibility. Similarly, Australia's Disability Discrimination Act (DDA) mandates that commercial and public entities must ensure their digital services, including AI tools, are accessible to everyone. These initiatives reflect a growing recognition of the global need for standardized accessibility practices in AI.

International collaboration is crucial in fostering a harmonized approach to accessibility standards. Various transnational initiatives are working to create unified frameworks that transcend borders. Initiatives like the Global Initiative for Inclusive Information and Communication Technologies (G3ict) and the International Telecommunication Union (ITU) focus on promoting and developing ICT accessibility. These bodies work toward creating guidelines and standards that can be adopted universally, facilitating the deployment of accessible AI technologies across different jurisdictions.

One of the critical challenges in implementing these international standards and practices is the need for localized adaptation. While international guidelines provide a foundational framework, they must be tailored to meet local needs, cultural contexts, and existing infrastructure. This requires ongoing dialogue between international bodies, local governments, industry stakeholders, and the disability community to ensure that the adopted standards are both globally coherent and locally relevant.

Countries with emerging economies often face additional hurdles in adopting and enforcing these standards. Limited financial resources, lack of technical expertise, and inadequate infrastructure can hamper efforts to implement accessible AI technologies. To address these disparities, international aid and collaboration can play a vital role. Programs that offer financial assistance, technical training, and infrastructure development support can help bridge the gap, ensuring that the benefits of accessible AI technologies are realized globally.

Moreover, the rapid pace of AI innovation presents a unique challenge to standardization efforts. The dynamic nature of AI technologies means that standards must be continually updated to keep pace with new developments. This requires a proactive approach from standards organizations, involving constant research, stakeholder engagement, and iterative revisions to ensure that guidelines remain relevant and effective.

Ethical considerations are also integral to the establishment and practice of international standards in AI. Ethical guidelines must ensure that AI technologies uphold principles of fairness, equity, and respect for human dignity. This includes addressing biases in AI algorithms, ensuring transparent and accountable AI governance, and prioritizing user consent and data privacy. International ethical standards, such as those proposed by the Institute of Electrical and Electronics Engineers (IEEE) and the Organisation for Economic Co-operation and Development (OECD), provide valuable frameworks for integrating these ethical imperatives into AI development.

Practically speaking, the implementation of international standards and practices involves a multi-faceted approach. It begins with rigorous research and development to ensure that AI technologies meet required accessibility standards. This is followed by comprehensive testing and validation to confirm compliance. Lastly, continuous monitoring and evaluation are essential to identify and address any gaps or issues that arise post-deployment. Engaging users with disabilities in each stage of this process is crucial, as their insights and feedback provide invaluable perspectives on usability and effectiveness.

The global movement towards standardizing AI accessibility is a testament to the potential of technology to bridge divides and enhance the quality of life for people with disabilities. As countries and organizations continue to collaborate, share knowledge, and innovate, they pave the way for a more inclusive future where AI technologies are accessible and beneficial to all.

In conclusion, international standards and practices form the bedrock of developing accessible AI technologies on a global scale. Through the combined efforts of international organizations, national governments, and local communities, significant strides are being made towards realizing a world where AI empowers individuals with disabilities. As we continue to advance in this field, the ethical, practical, and collaborative dimensions of standardization will remain pivotal in ensuring that the promise of AI is fulfilled for everyone.

Chapter 20:
Collaboration Between Stakeholders

Collaboration between stakeholders is essential for the successful development and deployment of AI technologies that cater to people with disabilities. This collaboration involves an interdisciplinary approach where technologists, healthcare professionals, educators, policy makers, and individuals with disabilities work in concert to ensure that the solutions developed are truly inclusive. Partnerships in development serve to pool resources, expertise, and perspectives, leading to innovative and practical solutions. The roles of government and NGOs are equally crucial, providing the necessary frameworks, funding, and advocacy to support these endeavors. By bridging the gap between various sectors, these collaborations not only drive technological advancement but also ensure that the societal and ethical implications are meticulously considered, ultimately enriching the lives of individuals with disabilities through effective and empathetic AI applications.

Partnerships in Development

When it comes to developing AI technologies that can significantly enhance the lives of people with disabilities, collaboration is key. Partnerships in development are vital for bringing diverse perspectives and expertise to the table, ensuring that the solutions created are not only innovative but also user-centered and practical. These partnerships often involve a variety of stakeholders, including tech companies, academic institutions, non-profit organizations, healthcare providers, and, crucially, individuals with disabilities themselves.

In these collaborative efforts, technology companies play a pivotal role. Their technical know-how and resources are indispensable for turning conceptual ideas into workable AI solutions. Large tech firms like Google, Microsoft, and IBM have already shown what's possible through their extensive research and development in AI. However, the contributions of smaller startups can't be underestimated. Startups often bring fresh, out-of-the-box thinking and agile methodologies to the table, which can drive rapid innovation in accessibility technologies.

Academic institutions contribute through rigorous research and testing, ensuring that AI technologies meet high standards of efficacy and safety. Universities often run specialized research labs focusing on AI and accessibility, producing cutting-edge research that the industry can build upon. These labs also serve as incubators for tomorrow's innovators, nurturing the next generation of experts who will continue to push the boundaries of what AI can achieve for people with disabilities.

Non-profit organizations add another crucial dimension to these partnerships. They provide invaluable insights into the real-world challenges people with disabilities face. Organizations like the American Foundation for the Blind (AFB) and the National Association of the Deaf (NAD) have been instrumental in guiding tech companies to develop more accessible and inclusive solutions. These organizations often serve as bridges between the tech community and end-users, ensuring that the technologies being developed are both needed and usable.

Healthcare providers bring a medical perspective to these collaborations. They understand the clinical and therapeutic needs of people with disabilities and can help shape AI technologies to meet those needs effectively. Whether it's developing AI systems for early diagnosis and intervention or creating tools for ongoing support and rehabilitation, the involvement of healthcare professionals ensures that the solutions are not just technologically advanced but also medically sound.

Individuals with disabilities are perhaps the most important partners in this collaborative ecosystem. They offer lived experiences and firsthand knowledge that can guide the development process, ensuring that the end products are truly beneficial. By involving them in every stage—from ideation and prototyping to testing and feedback—developers can create solutions that genuinely enhance accessibility and independence. This user-centered approach not only leads to more effective products but also empowers individuals with disabilities by giving them a voice in the creation of technologies that affect their lives.

One shining example of successful collaboration is the development of smart assistants that cater to diverse needs. Tech giants have teamed up with non-profits and academic institutions to create virtual assistants that can help individuals with disabilities manage their daily activities more effortlessly. By integrating voice commands, tactile feedback, and visual recognition technologies, these smart assistants take into account various types of disabilities, making them versatile tools for enhancing daily living.

The role of government and public sector organizations can't be ignored in these partnerships. They often provide the regulatory framework and funding opportunities that make large-scale projects feasible. Governments can foster innovation by offering grants, tax incentives, and other financial supports aimed at developing accessible technologies. Policies that mandate accessibility standards also push tech companies to prioritize inclusive design in their development processes.

International collaborations add yet another layer of richness to these partnerships. Working with global organizations broadens the scope of research and development, incorporating a variety of cultural perspectives and needs. This is especially important for creating AI solutions that are adaptable and relevant across different contexts and regions. It also opens up channels for sharing best practices and leveraging international expertise in accessibility.

Even venture capitalists and angel investors are recognizing the potential impact of investing in AI technologies for accessibility. These financial backers bring with them not just funding, but also valuable business acumen and market strategies that can help bring products from prototype to market. Their involvement signals a growing recognition of the market potential for accessible technologies, which can encourage further investment and innovation in this space.

The synergy between these varied stakeholders often leads to the creation of holistic solutions that address multiple facets of accessibility and inclusion. By working together, these partners can pool their resources, share knowledge, and overcome challenges more effectively than they could alone. The cross-pollination of ideas and expertise leads to richer, more robust AI solutions that can make a tangible difference in people's lives.

For instance, consider the development of AI-powered communication aids for individuals with speech impairments. This kind of innovation requires input from speech language pathologists, AI experts, software developers, and end-users. Each group brings its own unique expertise, and without this collaborative effort, the final product would likely be less effective.

Another noteworthy example is the intersection of AI and education for children with cognitive disabilities. Here, educators, cognitive scientists, AI researchers, and caregivers must work together to develop personalized learning environments. These partnerships ensure that the AI tools developed are pedagogically sound, technologically advanced, and tailored to meet the unique needs of each child.

To sum up, partnerships in development are the backbone of creating effective AI solutions for enhancing accessibility and independence for people with disabilities. By leveraging the strengths and expertise of diverse stakeholders, we can ensure that these technologies are not only innovative but also inclusively designed and practically applicable. This collaborative approach enriches the development process, leading to more impactful and sustainable solutions that can make a real difference in people's lives.

Roles of Government and NGOs

When it comes to leveraging artificial intelligence (AI) to improve the lives of people with disabilities, the roles of government and non-governmental organizations (NGOs) cannot be overstated. Collaboration between these stakeholders is pivotal for creating a world where AI-driven accessibility solutions are not just innovative but also inclusive and widely accessible.

Governments have the unique ability to legislate and regulate, which sets the stage for a conducive environment that fosters innovation while ensuring equity. By formulating policies that encourage research and development in AI, governments can offer grants and tax incentives to tech companies and academic institutions engaged in creating assistive technologies. For instance, earmarking funds specifically for AI projects aimed at disability inclusion can significantly expedite progress in this field.

Moreover, governments can play an essential role in developing guidelines and standards that ensure AI applications are accessible and inclusive from the outset. Creating robust frameworks for ethical AI development and mandating adherence to these guidelines across various industries can ensure that the interests and needs of people with disabilities are always considered. For example, standards that require AI systems to be compatible with screen readers or voice command functionalities can go a long way in enhancing accessibility.

Infrastructure is another critical area where government participation is crucial. Public funding can be directed towards building and maintaining the necessary infrastructure—such as high-speed internet and public tech labs—that enables the widespread use and testing of AI-powered assistive technologies. Without the requisite infrastructure, even the most groundbreaking innovations can remain out of reach for many, particularly in rural or underdeveloped areas.

On the other hand, NGOs often serve as the voice for the marginalized and underrepresented. These organizations can highlight the challenges faced by people with disabilities and advocate for solutions by collaborating with tech companies, governments, and educational institutions. Because NGOs are usually well-versed in the unique needs and nuances of various disability communities, they are invaluable partners in the design and deployment of AI solutions. They bring to the table an in-depth understanding of user requirements, which helps in creating more intuitive and effective assistive technologies.

NGOs can also act as intermediaries that bridge the gap between the technical experts developing AI technologies and the people who will eventually use them. Through workshops, focus groups, and pilot programs, NGOs can provide critical insights into user experiences and feedback, which can be used to refine and improve AI applications. Furthermore, they can facilitate training programs to educate both users and practitioners on the effective use of AI tools, ensuring that these technologies are utilized to their full potential.

Besides advocacy and training, NGOs can play a crucial role in monitoring the implementation and impact of AI technologies. They can undertake independent studies to evaluate how these technologies are performing in real-world scenarios and whether they are genuinely enhancing the quality of life for people with disabilities. This kind of accountability ensures that the focus remains not just on technological innovation but also on tangible, positive outcomes.

Another critical role NGOs can play is in promoting awareness and understanding of AI technologies among the general public and within communities of people with disabilities. Public awareness campaigns, community workshops, and informational webinars can help demystify AI and its applications, thereby reducing apprehensions and encouraging more widespread adoption. When people understand how AI can be a tool for empowerment, there's likely to be more grassroots support for its deployment, which can spur further innovation and acceptance.

The synergy between governments and NGOs is crucial for identifying gaps in current practices and ensuring that comprehensive, multi-faceted solutions are developed. While governments can wield the power of policy and resources, NGOs bring the on-the-ground insights and user-focused perspectives that are essential for creating truly inclusive AI solutions. Both have roles that complement each other, and their collaboration often results in more robust, effective, and sustainable outcomes.

For instance, consider the development and deployment of AI-driven hearing aids. Governments can provide funding and tax incentives, set standards for quality and accessibility, and ensure that the necessary technological infrastructure is in place. Simultaneously, NGOs can work with affected communities to gather user insights, advocate for the needs of people with hearing impairments, and run campaigns to raise awareness about the availability and benefits of these new technologies.

Furthermore, NGOs can assist governments by identifying international best practices and advocating for their adoption. Through their global networks, many NGOs can offer insights into what's working well in other parts of the world and propose how these practices can be adapted locally. Sharing knowledge across borders helps to accelerate progress and ensures that no one is left behind.

Lastly, NGOs can play a pivotal role in the long-term sustainability of AI initiatives aimed at assisting people with disabilities. They can help to create community ownership of projects, ensuring that they are not just top-down implementations but are instead co-created with input from the people who will benefit the most. This participative approach helps to ensure that assistive technologies are culturally appropriate, user-friendly, and able to evolve in response to changing needs and conditions.

In conclusion, the roles of government and NGOs in leveraging AI for the benefit of people with disabilities are both crucial and complementary. Through legislation, funding, advocacy, training, and community engagement, these stakeholders can work together to make sure that AI technologies reach their full potential as tools of empowerment and inclusion. Their combined efforts ensure that advancements in AI not only offer cutting-edge solutions but also align with the real-world needs and aspirations of people with disabilities.

Chapter 21:
Funding and Support for AI Projects

Securing funding and support for AI projects dedicated to improving the lives of individuals with disabilities is a multifaceted endeavor that requires navigating various sources of financial backing and incentives. Public grants, private investments, and non-profit organizations often play vital roles in this landscape. Governments and international bodies offer grants and tax incentives aimed at fostering innovation in accessibility technologies. Private sector investments from tech giants and venture capitalists can provide the crucial financial lifeline needed to transform visionary concepts into tangible solutions. Additionally, crowd-funding platforms and philanthropic efforts channel community enthusiasm into financial support, creating a groundswell of grassroots endorsement. Understanding the diverse array of funding opportunities, from traditional grants to modern crowdfunding, is essential for driving forward impactful, inclusive AI projects that break barriers and enhance the quality of life for those with disabilities.

Sources of Funding

When embarking on AI projects aimed at enhancing the lives of people with disabilities, securing adequate funding is crucial. In this chapter, we explore the multiple avenues through which such projects can obtain financial support, emphasizing the various stakeholders and mechanisms that make these ambitious initiatives possible. By comprehensively understanding the funding landscape, anyone interested in driving AI accessibility projects can align their efforts strategically and sustainably.

Firstly, the most straightforward source of funding often comes from **government grants**. Many national and local governments recognize the transformative potential of artificial intelligence in the realm of accessibility and inclusivity. As a result, they offer grants specifically targeted toward technology innovation for marginalized groups. Programs like the U.S. Department of Health and Human Services' Small Business Innovation Research (SBIR) grants and the European Union's Horizon Europe framework provide substantial resources to innovators aiming to leverage AI for social good. Government grants typically have specific application processes, criteria, and objectives, but they provide substantial economic backing with fewer demands for equity or ownership.

Another significant source of funding is through **non-profit organizations** and **charities**. A multitude of foundations focus on disability rights, technological advancements, and social equity. Entities such as the Bill & Melinda Gates Foundation, the Wellcome Trust, and the Ford Foundation have funded numerous AI initiatives designed to benefit those with disabilities. These organizations often focus on the alignment of their mission statements with the impact goals of the project, emphasizing the ethical implications and societal benefits.

Corporate sponsorship and *partnerships* also play a pivotal role in supporting AI projects. Large technology companies, including Google, Microsoft, and IBM, have dedicated accessibility funds and innovation labs. These corporations often seek collaborative opportunities that enable them to advance their own technology while contributing to the broader societal cause. Through programs like Microsoft's AI for Accessibility, tech giants provide not just financial resources but also technical support, mentorship, and access to cutting-edge tools. Such partnerships can be mutually beneficial, fostering innovation that aligns with the strategic objectives of both parties involved.

Furthermore, **venture capital (VC)** firms are increasingly interested in funding AI solutions that promise a social impact. Although traditionally profit-driven, many VCs now house dedicated funds for social innovation. Impact investors, in particular, look for ventures that offer substantial returns both financially and socially. Angel investors and specialized venture funds like the Social Venture Fund and the Rise Fund are examples of investors willing to take on cutting-edge projects aimed at inclusivity. It's crucial for AI developers to craft compelling pitches that highlight both the market potential and the societal impact of their innovations when seeking VC funding.

We must also acknowledge the expanding role of *crowdfunding* platforms. Websites like Kickstarter, Indiegogo, and GoFundMe have democratized access to capital, allowing innovators to pitch their projects directly to the public. These platforms can be particularly effective in raising awareness and generating initial funding for smaller-scale or exploratory projects. Crowdfunding serves a dual purpose: securing early-stage capital and validating the market interest and social value of the proposed AI solution.

Academic institutions and **research grants** constitute another viable funding pathway. Universities and specialized research bodies often have grants aimed at fostering innovation in fields that overlap with artificial intelligence and disability support. Institutions like the National Science Foundation (NSF) and various university-affiliated research centers offer grants for projects that push the boundaries of current knowledge and technology. Academic partnerships not only provide funding but also engage students and faculty in research, ensuring a rigorous approach to data collection, validation, and outcomes assessment.

Lastly, *philanthropic crowdfunding* platforms and social enterprises offer unique funding challenges and opportunities. Entities that operate at the intersection of philanthropy and business, like Kiva and the Acumen Fund, often support ventures that drive significant social change. By leveraging peer-to-peer lending models or social impact bonds, these platforms bring in loans or investment with an expectation of both financial and social returns. This model encourages sustainable business practices while emphasizing impactful outcomes.

For long-term sustainability, it's equally important to consider **fundraising campaigns** and **donor engagement**. Building a robust network of individual donors who are passionate about accessibility and AI can provide a steady flow of donations to support incremental and ongoing costs. Campaigns can include annual drives, gala events, or targeted outreach to high-net-worth individuals. Effective donor engagement requires consistent communication, transparency in how funds are used, and demonstrable impact, often through success stories and case studies.

Another emerging funding avenue worth exploring is **token sales** via blockchain technology. While still a nascent area, some projects have utilized initial coin offerings (ICOs) or security token offerings (STOs) to raise capital. This method involves issuing tokens or coins on a blockchain, providing backers with a stake in the project's future success. While this avenue comes with its own set of regulatory and technical challenges, it holds potential for tech-savvy AI projects aiming to harness decentralized funding sources.

In summary, funding AI projects geared toward improving accessibility for people with disabilities requires a multi-pronged approach. Government grants, non-profit funding, corporate partnerships, venture capital, crowdfunding, academic support, philanthropic crowdfunding, donor engagement, and emerging technologies—all play essential roles. By diversifying funding sources, project developers can not only secure the necessary capital but also build a resilient and dynamic base of support. This comprehensive strategy ensures that projects can sustain their momentum and continue to innovate, ultimately achieving their goal of creating a more inclusive and empowering world for people with disabilities.

Grants and Incentives

Creating an environment where artificial intelligence (AI) can flourish, especially in the realm of accessibility for individuals with disabilities, requires substantial funding and inventive incentive programs. Numerous organizations recognize the transformative potential of AI and have created extensive grants and incentives to foster innovation. Understanding and tapping into these resources can make a meaningful difference in the development and deployment of AI technologies aimed at enhancing the lives of people with disabilities.

Governments worldwide understand that AI can reshape almost every sector, from healthcare to education, and significantly improve the quality of life for individuals with disabilities. As a result, many have initiated grant programs to stimulate research, development, and implementation of AI-driven assistive technologies. These grants often focus on areas such as improving accessibility, increasing independence, and enhancing communication for people with varying disabilities.

For example, in the United States, the National Institute on Disability, Independent Living, and Rehabilitation Research (NIDILRR) provides funding opportunities through various grant programs targeted specifically at advancing technology for people with disabilities. The NIDILRR's Small Business Innovation Research

(SBIR) program is particularly aimed at supporting innovative research and development by small businesses that can lead to commercial products benefiting individuals with disabilities.

Similarly, programs like the European Commission's Horizon 2020 have a strong emphasis on inclusive technologies. Horizon 2020 offers substantial financial support for projects that aim to use AI for social good, including improving the lives of those with disabilities. These programs typically encourage consortia of companies, research institutions, and nonprofit organizations to collaborate on cutting-edge AI projects.

Incentives aren't limited to direct grants alone. Tax incentives also play a crucial role in encouraging organizations to invest in AI projects. For instance, many countries offer tax credits or deductions for research and development activities, which can significantly lower the financial burden of developing new AI technologies. Such incentives make it more feasible for smaller companies and startups to venture into the development of assistive technologies without facing insurmountable financial risks.

Private foundations and philanthropic organizations also contribute significantly to funding AI projects tailored for disability support. Organizations such as the Ford Foundation and the Bill and Melinda Gates Foundation have been known to fund initiatives that leverage technology to solve social issues including accessibility. These foundations often host innovation challenges or grant competitions, incentivizing researchers and developers to create novel solutions for barriers faced by people with disabilities.

Corporate social responsibility (CSR) initiatives are another avenue through which significant funding and support can be garnered. Tech giants like Google, Microsoft, and IBM offer grant programs and funding opportunities aimed at leveraging AI for social impact. These companies not only provide financial aid but often also lend their expertise and technological infrastructure to support these projects. For instance, Microsoft's AI for Accessibility grant program offers grants and support to those developing AI solutions that help to address the challenges faced by people with disabilities globally.

Incubators and accelerators focused on social impact and healthcare innovation offer another layer of support. These platforms often provide early-stage funding, mentorship, and networking opportunities to startups and researchers working on AI-based assistive technologies. Programs like Y Combinator, Techstars, and MassChallenge have dedicated tracks for health and accessibility, ensuring that meaningful projects receive the practical support they need to grow and scale.

Beyond financial support, grants and incentives often come with additional perks that can be invaluable to the success of AI projects. These can include access to specialized research labs, data sets, and computing resources that would otherwise be too costly for small teams. Moreover, many grants come with stipulations for collaboration, encouraging partnerships with academic institutions, healthcare providers, and disability advocacy groups.

Another vital aspect of these programs is their focus on sustainability. Sustainable funding models ensure that projects don't just launch but can also maintain and scale their operations over time. Some grant programs offer multi-year funding or opportunities to renew grants based on milestones and progress, providing a stable financial runway for extended research and development phases.

It's not all about large-scale projects either. Microgrants and seed funding are increasingly available for smaller initiatives or pilot projects. These smaller financial boosts are often the first step in validating an idea and proving its viability before seeking larger-scale funding. Organizations like the Disability Rights Fund provide such microgrants, enabling grassroots initiatives to get off the ground.

In summary, the landscape of grants and incentives for AI projects aimed at assisting those with disabilities is robust and varied. Between government programs, private foundations, corporate grants, tax incentives, and incubators, there are myriad opportunities to secure the necessary financial support to bring innovative AI solutions to life. Leveraging these resources effectively can catalyze significant advancements, breaking down barriers and creating new possibilities for individuals with disabilities. The impact of these efforts can be profound, offering enhanced accessibility, increased independence, and a significantly improved quality of life for the disabled community.

Chapter 22:
Accessibility Standards and Guidelines

Establishing and adhering to accessibility standards and guidelines is crucial in ensuring that artificial intelligence technologies genuinely benefit individuals with disabilities. These standards serve as foundational principles that guide the development and implementation of inclusive AI solutions. Governments, organizations, and developers must collaborate to adopt and integrate these guidelines effectively, thereby enhancing the usability and accessibility of technology. From web accessibility standards like WCAG to device-specific protocols, these frameworks ensure that technological advancements do not leave anyone behind. Proper implementation of these best practices not only empowers those with disabilities to lead more independent lives but also fosters an inclusive society where technology acts as an equalizer. By committing to these standards, we're not just following a set of rules; we're advocating for a world where technology serves as a bridge rather than a barrier.

Current Standards

The landscape of accessibility standards is continuously evolving to keep pace with technological advancements and societal needs. In the realm of AI technologies, these standards are critical to ensuring that innovations genuinely enhance the lives of people with disabilities. Current standards aim to provide frameworks that guide the development and implementation of accessible technologies, ensuring they are equitable, functional, and user-friendly.

One of the most foundational standards is the Web Content Accessibility Guidelines (WCAG), developed by the World Wide Web Consortium (W3C). These guidelines are essential for digital content providers and developers to ensure that web applications and sites are accessible to all users, including those with disabilities. WCAG provides a comprehensive set of recommendations designed around four principles: perceivable, operable, understandable, and robust, often abbreviated as POUR. These principles ensure that users of all abilities can access, navigate, and interact with web content effectively.

In addition to WCAG, the Americans with Disabilities Act (ADA) also plays a pivotal role in shaping accessibility standards, particularly within the United States. While originally focused on physical spaces, the ADA has been interpreted to apply to digital environments, ensuring that websites and online services are accessible to individuals with disabilities. This has significant implications for AI technologies, as these systems often rely on web-based interfaces or digital platforms.

Internationally, the ISO/IEC 40500:2012 standard, which is essentially the international adaptation of WCAG 2.0, provides a globally recognized framework for accessibility. It aligns with other regional standards and ensures a harmonized approach to creating accessible technologies worldwide. This standard is vital for developers working on global AI solutions, as it helps maintain consistency and quality across different markets and user bases.

Moreover, the Section 508 standards of the Rehabilitation Act in the United States mandate that federal agencies' electronic and information technology is accessible to people with disabilities. These standards are updated periodically to include the latest technological advancements, ensuring that innovations are always within reach of all users, regardless of their abilities. For AI technologies, Section 508 compliance is crucial, particularly for products and services intended for use by government agencies.

In the context of mobile technologies, the Mobile Accessibility Guidelines provide specific requirements to ensure that mobile applications are usable by people with disabilities. These guidelines address issues unique to mobile devices, such as touch navigation and voice commands, which are critically relevant for the growing number of AI applications on smartphones and tablets. Ensuring accessibility in mobile AI applications enhances independence and mobility for users, making everyday tasks more manageable and efficient.

Another key standard is the Accessible Rich Internet Applications (ARIA) specification, which focuses on making web content and web applications more accessible to people with disabilities. ARIA is particularly important for dynamic web content and applications, which are characteristic of many AI-driven technologies. By providing additional information to assistive technologies, ARIA allows for more interactive and complex functionalities to be accessible, preserving the richness of user experiences for everyone.

Compliance with these standards is not just about legal adherence; it's about fostering inclusivity and equity in technological development. Developers and organizations must prioritize accessibility as a core principle, integrating it into the early stages of design and development processes. This user-centered approach ensures that products are intuitively accessible and effective for all potential users.

To truly capture the spirit of current accessibility standards, it's essential to consider the perspective of Universal Design—a philosophy that supports creating products and environments usable by everyone, regardless of age, ability, or status. Universal Design principles overlap significantly with accessibility standards, advocating for simplicity, flexibility in use, and perceptible information, all of which are paramount in AI technology development.

Moreover, ongoing research and collaboration in the field of accessibility standards are crucial. Stakeholders from various domains—developers, educators, healthcare professionals, and people with disabilities—must work together to continually refine and update standards. This collaborative effort ensures that accessibility guidelines stay relevant and comprehensive, addressing emerging technologies and evolving user needs.

However, current standards are not without their challenges. One major issue is the rapid pace of technological advancement. As AI technologies evolve, there can be a lag in updating and establishing new standards, leading to periods where certain innovations might not be fully accessible. Continuous dialogue and proactive updates from standard-setting bodies are necessary to bridge this gap.

Another challenge lies in the interpretations of standards across different industries and organizations. While standards like WCAG and ISO/IEC 40500:2012 provide strong frameworks, their application can vary, leading to inconsistencies in accessibility across different platforms and technologies. Clear guidance, education, and support for developers and organizations are essential to ensure uniform implementation and compliance.

In the coming years, we can expect to see further refinement and expansion of accessibility standards to encompass new dimensions of AI technologies. With the advent of more sophisticated AI systems, including machine learning and neural networks, standards will need to address how these technologies can be made transparent, explainable, and user-inclusive. Ensuring that AI systems can be audited and understood by users with varying cognitive capabilities will be a significant focus.

Ultimately, current accessibility standards provide a blueprint for creating inclusive, equitable, and user-friendly AI technologies. By adhering to these guidelines, developers have the opportunity not only to comply with legal requirements but to make meaningful impacts on the lives of people with disabilities. These standards serve as a reminder that innovation must go hand-in-hand with inclusivity—ensuring that the benefits of AI are shared equitably across all segments of society.

Implementing Best Practices

When it comes to accessibility standards and guidelines, implementing best practices isn't merely a technical requirement—it's an ethical imperative and a gateway to meaningful inclusion. AI technologies present a powerful toolset to enhance accessibility for people with disabilities. However, the effectiveness of these technologies hinges on adhering to best practices, ensuring they are both universally applicable and individually adaptable.

First and foremost, accessibility must be a consideration from the very inception of any AI-driven project. Too often, accessibility features are tacked on as an afterthought—if they are included at all. This approach isn't just inefficient; it's exclusionary. By integrating accessibility standards from the planning phase, developers set the groundwork for a more inclusive technology landscape. Utilizing established frameworks such as the Web Content Accessibility Guidelines (WCAG) and the Accessible Rich Internet Applications (ARIA) provides a robust starting point for building inclusive technologies.

One crucial aspect of implementing best practices is user involvement. Involving people with disabilities in the design and development stages ensures that the technology is genuinely meeting their needs. Not only does this help identify potential oversights or issues early on, but it also fosters a sense of ownership and relevance among users. A collaborative design process can include various methods such as user surveys, focus groups, and usability testing sessions that involve a diverse group of participants.

Moreover, adopting a user-centered design approach promotes adaptability and personalization. AI technologies can and should be tailored to individual needs and preferences. For instance, customizable interfaces can accommodate different types of disabilities by allowing users to adjust settings such as text size, color contrast, and voice command sensitivity. Machine learning algorithms can further enhance this by learning the user's preferences over time, making the technology more intuitive and effective.

Automation plays a pivotal role in implementing accessibility best practices within AI systems. For example, automated tools can be used to scan and identify accessibility issues in code, which ensures that standards are consistently met. However, it's vital to remember that automation should supplement, not replace, human oversight. Developers must perform manual checks and user testing to validate these automated efforts, ensuring that they meet the nuanced needs of people with disabilities.

Regular training and education are just as important for maintaining best practices. Developers, designers, and all team members should undergo accessibility training to stay updated with the latest standards and innovations. This training can cover a range of topics, from understanding different disabilities to learning how to apply accessibility guidelines in their work. Continuous education fosters an environment where accessibility becomes an intrinsic part of the company's culture, rather than a box-ticking exercise.

Documentation and transparency are also critical components. Comprehensive documentation ensures that all team members have access to the same information and understand the rationale behind certain accessibility choices. Transparency, on the other hand, builds trust with users who know that their needs and feedback are being taken seriously. Openly sharing the methodologies, challenges, and successes related to accessibility efforts can inspire other developers to adopt similar best practices.

Flexibility and scalability should be core principles when implementing best practices in AI accessibility. The rapid evolution of both AI technologies and accessibility standards necessitates systems that can quickly adapt to new requirements. Modular design solutions enable easier updates and modifications, ensuring that technology remains accessible as it evolves. For example, AI-driven assistive technologies can be built on scalable platforms that allow for seamless integration of new features and improvements based on user feedback and technological advancements.

Additionally, fostering a community of practice around accessibility can have profound benefits. Establishing forums, working groups, and partnerships with organizations that focus on disability advocacy can provide valuable insights and support. These collaborations can accelerate the adoption of best practices by sharing knowledge, resources, and success stories. Engaging with the broader community also amplifies the impact of your work, highlighting the social and collective benefits of accessible technology.

Ethical considerations should underpin every aspect of AI development for accessibility. This involves not only following legal frameworks and standards but also engaging with the ethical dimensions of AI use. Questions surrounding data privacy, consent, and the potential for bias in AI algorithms must be addressed transparently. Developers should implement robust privacy safeguards and strive for fairness and inclusivity in their algorithmic decisions, ensuring that AI benefits all users equitably.

The landscape of accessibility is ever-changing, driven by both technological advancements and evolving societal expectations. Staying ahead of these changes requires a proactive approach. Regularly reviewing and updating accessibility protocols, staying informed about new developments, and being open to innovative practices are essential. By doing so, developers can not only comply with the latest standards but also push the boundaries of what is possible, creating more inclusive and empowering technologies.

Implementing best practices in accessibility is a continuous journey rather than a fixed destination. It requires ongoing commitment, investment, and a willingness to learn and adapt. However, the rewards are immense. By prioritizing accessibility, we not only create better technologies but also foster a more inclusive and empathetic society. In this endeavor, AI has the potential to be a transformative force, breaking down barriers and opening up new opportunities for people with disabilities worldwide.

As we move forward, it is vital to maintain a balanced perspective—celebrating our achievements while recognizing the work still to be done. Implementing best practices in accessibility is about more than compliance; it's about transforming lives and creating a world where everyone has the opportunity to thrive. By adhering to these principles, we can ensure that AI technologies serve as a force for good, enhancing accessibility and, by extension, the quality of life for all individuals.

Chapter 23: Case Studies of Successful AI Applications

In exploring the real-world impacts of AI on individuals with disabilities, several standout examples illustrate the transformative power of these technologies. For instance, consider how a sophisticated AI-driven visual recognition tool has empowered visually impaired individuals by converting visual information into audible descriptions in real time, significantly enhancing their independence. Another exemplar is the implementation of AI-powered speech recognition systems that have revolutionized communication for those with speech impairments, enabling them to interact more fluidly and effectively in both personal and professional settings. Each case study not only demonstrates a clear improvement in quality of life for individuals but also provides invaluable lessons on the scalable and adaptable nature of AI solutions. These success stories highlight that, when harnessed thoughtfully, AI can create more inclusive societies where technological innovation truly serves as an enabler of empowerment and autonomy.

Detailed Examples

To truly understand the transformative power of AI in enhancing the lives of people with disabilities, we need to look at specific, real-world examples where AI has been successfully implemented. These cases show the tangible impact of technology, moving beyond theory and into practical application.

Consider the groundbreaking application of AI in visual impairment solutions. One of the most notable examples is *Seeing AI*, a mobile app developed by Microsoft. This app describes the world around its users through the power of artificial vision. It can read text aloud, describe scenes, identify product barcodes, and even recognize faces. For someone who is visually impaired, this tool offers a level of independence that was previously unimaginable. They can navigate their environment more confidently and perform everyday tasks that most of us take for granted.

Another inspiring example comes from the realm of speech and communication disabilities. The AI-powered communication device *Proloquo2Go* is designed to give a voice to those who cannot speak. This app, available on iPads and other devices, uses natural language processing to turn typed text into spoken words. It's highly customizable, catering to a wide range of motor and cognitive abilities. For many users, this app transforms their ability to interact with the world, allowing them to express themselves in ways they previously couldn't.

The success of AI applications isn't just limited to mobile apps. In education, AI-based platforms like *Smart Sparrow* offer personalized learning experiences to students with cognitive disabilities. These platforms adapt the content based on the user's performance, ensuring that they receive the support they need to learn effectively. By creating a tailored educational journey, these AI tools help bridge the educational gap and foster a more inclusive learning environment.

In healthcare, *AI-based diagnostic tools* are revolutionizing the way medical professionals identify and treat conditions. For instance, IBM's Watson for Health uses AI algorithms to analyze vast amounts of medical data, helping doctors diagnose and treat diseases more accurately and quickly. For individuals with disabilities, timely and accurate diagnosis can mean the difference between manageable conditions and severe complications. These AI systems provide a second pair of eyes, ensuring that nothing slips through the cracks.

Mobility is another area where AI has made remarkable strides. Autonomous vehicles equipped with AI technologies are being engineered to assist people with physical disabilities. Companies like Waymo are testing self-driving cars that can be summoned with an app, eliminating the need for physical driving capabilities. This technology promises to provide unprecedented freedom and independence to those who have mobility challenges, enabling them to commute without relying on others.

AI's role extends into emotional and mental support too. Virtual companions like Woebot, an AI-driven chatbot, offer therapeutic conversations to individuals struggling with mental health issues. Woebot uses cognitive-behavioral techniques to provide real-time assistance, monitoring users' emotional states and offering personalized coping strategies. For people who may not have access to traditional therapy, these AI tools provide a valuable alternative for emotional support.

Beyond individual use cases, entire communities also benefit from AI-driven accessibility improvements. The *Smart Cities Project* leverages AI to create urban environments that are inclusive and accessible. By integrating AI technologies into the infrastructure, these cities offer better navigation for individuals with disabilities. For example, AI-powered traffic signals equipped with audio cues and timing adjustments can help visually impaired individuals cross streets safely.

On the employment front, AI-powered tools like *JobAccess* are making significant contributions by assisting individuals with disabilities in finding and maintaining jobs. These platforms use AI to match job seekers' skills with suitable job openings, considering any accessibility requirements. They also provide training modules to help users develop necessary skills, break down employment barriers, and promote workplace inclusivity.

In robotics, companies like *ReWalk Robotics* have developed robotic exoskeletons that enable individuals with spinal cord injuries to walk again. These exoskeletons use AI to analyze the wearer's movements and provide the necessary support at the right time, effectively replacing lost function. It's a remarkable example of technology giving back autonomy and transforming the lives of those affected by severe mobility impairments.

In the realm of hearing impairments, AI plays a pivotal role too. Advanced hearing aids, such as those developed by Oticon, use machine learning algorithms to differentiate between background noise and speech. This selective amplification makes conversational comprehension significantly easier for the hearing impaired, improving their communication experiences. Such advancements have a profound impact, as they enhance social interaction and reduce the feeling of isolation often experienced by those with hearing challenges.

Digital accessibility is another significant area enriched by AI. Screen readers like NVDA and JAWS, which cater to the visually impaired, now incorporate AI to better understand and narrate web content. This intelligent processing goes beyond simply reading text, offering descriptions for images and other non-text content found online. With these tools, the internet becomes a more accessible place, opening up a world of information and opportunities to visually impaired users.

Artificial Intelligence isn't only transforming individual lives; it's also reshaping public policies and practices. Governments around the world are leveraging AI to enforce accessibility standards and improve public services. For instance, the use of AI-driven text analysis tools can ensure that public documents are accessible to everyone, including those with visual and cognitive impairments. By automating these processes, such tools help create an inclusive society where the rights and needs of people with disabilities are prioritized and met.

The beauty of AI applications lies in their potential to adapt and grow. As machine learning algorithms evolve, so do the capabilities of these assistive technologies. The future might hold even more sophisticated tools that offer personalized assistance in ways we can't fully imagine yet. This adaptation means continuous improvement and increased effectiveness in meeting the diverse needs of the disabled community.

It's also worth mentioning the synergies created through collaborative efforts between tech companies, healthcare providers, educational institutions, and advocacy groups. Projects like Google's Project Euphonia illustrate how collaborative efforts can yield groundbreaking results. By working together, stakeholders can pool their resources and expertise to develop AI tools that are more effective and widely accessible.

The critical takeaway from these detailed examples is that the integration of AI into the lives of those with disabilities isn't just a possibility—it's a reality that's already making a significant difference. These technologies do more than merely assist; they empower. They offer independence, enhance communication, open up educational and employment opportunities, and improve overall quality of life. AI is a beacon of hope, a testament to what can be achieved when human ingenuity meets technological innovation.

As we move forward, the importance of continued investment and innovation in these technologies cannot be overstated. It's essential to prioritize inclusivity in AI development and implementation, ensuring that the benefits of AI are accessible to everyone, regardless of their abilities. The journey of AI in disability support is just beginning, and the potential for future advancements is vast and incredibly promising.

Lessons Learned

Reflecting on the **Case Studies of Successful AI Applications**, it's clear that the journey toward integrating artificial intelligence to support those with disabilities has been both transformative and instructive. Through these case studies, several crucial lessons have emerged that can guide future efforts in this domain.

First and foremost, the importance of user-centered design cannot be overstated. Each successful AI application reviewed consistently involved users in the development process. By actively engaging users, developers ensured that the final products were not only functional but also met the real needs of individuals. This approach helped mitigate the risk of creating technologies that users find impractical or cumbersome.

Another significant lesson learned is the value of interdisciplinary collaboration. Successful projects often brought together experts from various fields including AI, healthcare, education, and the users themselves. This multifaceted approach allowed for a more holistic understanding of the challenges faced and led to more robust and innovative solutions. For instance, in developing AI-powered tools for the visually impaired, insights from ophthalmologists, software engineers, and visually impaired individuals themselves were all crucial.

Moreover, flexibility and adaptability emerged as critical factors. AI technologies that allowed for customization and personalization saw higher adoption rates. Disabilities vary widely in their manifestations and impacts; thus, solutions that could be tailored to individual needs proved most effective. This customizability not only improved usability but also fostered a sense of ownership among users, making them more likely to integrate these solutions into their daily lives.

The role of continuous feedback and iteration cannot be neglected either. Successful AI applications did not stop at development and deployment. Instead, they embraced a culture of ongoing improvement by constantly seeking user feedback and making necessary adjustments. This iterative process enabled developers to refine their technologies, ensuring they stayed relevant and effective in dynamic, real-world settings.

An equally important lesson is the potential of AI to enhance independence and self-sufficiency among users. Numerous case studies highlighted how AI-driven technologies empowered individuals to perform tasks that were previously difficult or impossible. Whether through smart assistants aiding communication or mobility solutions providing enhanced movement capabilities, the boost in confidence and independence was palpable. This not only improved the quality of life for users but also reduced the demand on caregivers and support systems.

Let's not overlook the significance of training and education. For AI applications to be truly successful in aiding those with disabilities, both users and support personnel need adequate training. Understanding the functionalities, as well as troubleshooting common issues, ensure that these tools are used to their full potential. Furthermore, educating the general public about these technologies fosters a more inclusive society where the benefits of AI are broadly understood and appreciated.

Many case studies also underscored the necessity of addressing ethical and privacy concerns from the outset. Trust is an integral factor in user adoption of AI technologies. Developers who proactively addressed data privacy, ethical implications, and inclusivity in their design and deployment phases saw greater acceptance of their solutions. Ensuring compliance with ethical standards and legal guidelines helped build trust and credibility, vital for long-term success.

Collaboration between stakeholders emerged as another cornerstone of success. Governments, NGOs, private companies, and educational institutions all played pivotal roles in the most impactful projects. Effective communication and coordinated efforts between these entities helped pool resources, share expertise, and align goals. This collaboration often facilitated access to funding, accelerated development timelines, and extended the reach of AI applications to underserved populations.

A deeper understanding of the diverse and multifaceted nature of disabilities also proved crucial. Technologies tailored to the unique challenges posed by different disabilities—whether visual, auditory, cognitive, or physical—tended to perform better. Recognizing and addressing the specific needs, limitations, and abilities of each user group enabled the creation of more effective and meaningful solutions.

Furthermore, addressing societal and cultural attitudes towards disability was essential. Successful applications didn't merely stop at technological advancements; they also incorporated educational components to shift public perceptions and increase acceptance of disability aids. By fostering a broader societal acceptance, AI applications had a more supportive environment to flourish.

Another area of focus was the economic aspects of developing and deploying AI technologies. Cost-effectiveness played a significant role in the sustainability of these solutions. Projects that managed to balance innovative features with affordability had a better chance of achieving widespread adoption. Partnerships that provided funding and incentives were also critical, ensuring that financial hurdles didn't impede technological progress.

Finally, a lesson that reverberates through these case studies is the transformative potential of AI when applied conscientiously and creatively. The AI applications chronicled here not only uplifted individual lives but also catalyzed broader social and systemic changes. They showcased the far-reaching impacts AI could have, extending beyond immediate user benefits to influence policy, education, and societal norms positively.

In conclusion, these lessons provide a robust roadmap for future endeavors in leveraging AI to support individuals with disabilities. By insisting on user-centered design, fostering interdisciplinary collaboration, encouraging flexibility and adaptability, and addressing ethical and privacy concerns, developers can create technologies that truly enhance accessibility and quality of life. The journey is ongoing, and these insights will undoubtedly pave the way for even more groundbreaking advancements in the future.

Chapter 24:
Building an Inclusive AI Ecosystem

Creating an inclusive AI ecosystem demands a multifaceted approach that marries innovative strategies with broad community involvement. We must foster a culture where inclusivity is prioritized from the ground up, embedding accessibility into every layer of AI development. This includes engaging with people with disabilities throughout the design and implementation stages to ensure technologies truly meet their needs. Building cross-sector partnerships can amplify these efforts, bringing together technologists, policymakers, and disability advocates to cultivate solutions that are equitable and practical. Ultimately, an inclusive AI ecosystem isn't just about accessibility—it's about empowering everyone to participate fully in society, thus driving forward a paradigm where technology serves as a universal enabler of independence and quality of life.

Strategies for Inclusivity

An inclusive AI ecosystem holds the promise of revolutionizing the way people with disabilities interact with technology, and by extension, the world around them. It's not just about developing advanced tools; it's about making sure these tools are accessible and beneficial to everyone, regardless of their abilities. But this goal requires deliberate and multifaceted strategies to ensure inclusivity from the ground up.

First and foremost, the development of AI technologies must integrate universal design principles. These principles advocate for products and environments to be inherently accessible to all people, to the greatest extent possible, without the need for adaptation. Essentially, this means building AI systems that are usable by the widest range of people, encompassing various disabilities and levels of technological proficiency.

One significant approach is engaging people with disabilities throughout the design and development process. Their firsthand experiences provide invaluable insights that developers might not otherwise consider. By including them in user testing, surveys, and even as part of the design team, the resulting AI solutions are far more likely to meet real-world needs and challenges.

Involving interdisciplinary teams can also drive inclusivity. Collaboration among AI developers, disability advocates, healthcare professionals, and educators can lead to more comprehensive and user-friendly technologies. Each group brings a unique perspective to the table, which can uncover potential issues and opportunities that a single-discipline team might overlook.

Developers should also prioritize adaptability in AI technology. This includes creating systems that can be customized to individual user preferences and needs. For instance, an AI-driven communication aid should allow modifications in speech patterns, vocabulary, and interface layout to accommodate various impairments. This flexibility ensures that users can tailor the technology to work for them, rather than struggling to adapt themselves to the technology.

Next, there's the importance of leveraging existing accessibility standards and guidelines. Organizations like the Web Content Accessibility Guidelines (WCAG) provide robust frameworks that AI technologies can integrate to ensure they cater to a wide audience.

Adhering to these standards helps in creating systems that are more universally accessible from the outset.

Moreover, ongoing education and training about the needs of people with disabilities are essential for developers and designers working on inclusive AI. Training sessions, workshops, and courses that focus on disability awareness can enlighten technologists about the diverse challenges and capabilities of potential users. This understanding can lead to more empathetic and effective design choices.

Policy also plays a critical role in fostering an inclusive AI ecosystem. Governments and organizations can enact regulations that mandate accessibility features in new AI technologies. Incentives, grants, and funding opportunities for inclusive projects can spur innovation and ensure that the development of AI for people with disabilities remains a priority.

Encouraging an open-source approach can further inclusivity. By making AI algorithms and software freely available, a larger community of developers can contribute to its enhancement, including individuals and organizations dedicated to accessibility. Open-source projects can benefit from the collective expertise and creativity of a global community, hastening the development of inclusive features.

It's also crucial to ensure AI literacy among people with disabilities. Providing accessible educational resources and training can empower these individuals to understand, use, and even contribute to AI technologies. When people with disabilities are not only users but also creators of AI, the ecosystem becomes truly inclusive.

Another strategic move is the deployment of continuous feedback loops. This means regularly seeking and incorporating feedback from users with disabilities throughout the product lifecycle. Regular updates and improvements based on user input help maintain relevance and effectiveness, ensuring that the technology evolves to meet changing needs.

Building partnerships with organizations that specialize in disability advocacy can amplify efforts towards inclusivity. These organizations have extensive knowledge and experience in addressing the needs and rights of people with disabilities. Collaborating with them can enrich the development and deployment processes of inclusive AI solutions.

Moreover, it's essential to address the digital divide. Although AI has the potential to significantly benefit people with disabilities, those without access to technology or reliable internet remain marginalized. Strategies must include efforts to improve digital literacy and access, ensuring that all individuals can benefit from AI innovations.

Creating inclusive AI is also about fostering an inclusive culture within AI development teams. A diverse workforce that includes individuals with disabilities can naturally lead to more inclusive outcomes. Diverse teams bring varied experiences and perspectives that can help avoid oversight or biases in the development of AI solutions.

Ultimately, achieving an inclusive AI ecosystem demands a commitment to ongoing evaluation and refinement. As both AI technologies and societal understandings of disability evolve, so too must the strategies for inclusion. This requires a willingness to remain flexible and make iterative improvements based on new findings and user feedback.

In closing, the journey to building an inclusive AI ecosystem is multifaceted and ongoing. It requires a commitment to universal design, inclusive participation, interdisciplinary collaboration, adaptability, adherence to standards, continuous education, policy support, open-source practices, AI literacy, regular user feedback, partnerships with advocacy groups, addressing the digital divide, fostering workplace diversity, and a culture of iterative improvement. By focusing on these strategies, we can ensure that AI technologies not only empower but also include people with disabilities at every step.

Community Involvement

Building an inclusive AI ecosystem isn't just a technical endeavor—it's a deeply social one. To truly harness the potential of AI in enhancing the lives of people with disabilities, we must actively involve the community at every step. The ingenuity and innovation behind impressive AI technologies often stem from understanding nuanced, real-world challenges. And there's no better source of this understanding than the people who live with disabilities themselves.

Community involvement begins with inclusive dialogue. Various stakeholders—including people with disabilities, their caregivers, educators, healthcare professionals, and tech developers—must come together to identify needs and brainstorm solutions. This isn't about token gestures but about genuine, ongoing conversations that respect and elevate the lived experiences of individuals with disabilities.

It's crucial to establish forums where people with disabilities can share their experiences and needs directly with developers and researchers. Such platforms could take many forms: town hall meetings, online discussion boards, focus groups, or even interactive workshops. These sessions should be designed to be accessible to all, making use of sign language interpreters, captioning services, and alternative formats to ensure full participation.

Establishing strong feedback loops is another vital aspect. Once AI solutions are developed, the community should be engaged in rigorous testing phases. User feedback can reveal unanticipated issues and provide insights for improvements. Incorporating iterative feedback helps not only in refining the technology but also in building trust between developers and the community. When individuals see their input valued and acted upon, they're more likely to feel a sense of ownership and actively participate in future projects.

Moreover, fostering partnerships with organizations that specialize in disability advocacy can amplify these efforts. These organizations often have a nuanced understanding of the specific challenges faced by their communities and can serve as valuable intermediaries. Collaborations with advocacy groups can also bring credibility and wider acceptance to new technologies.

Education and awareness campaigns are integral to community involvement. Many people, including those with disabilities, may be unaware of the AI tools available to them or may have misconceptions about their efficacy and safety. Awareness initiatives could range from informative sessions, webinars, and hands-on demonstrations, to integrating AI literacy into existing educational programs. This equips individuals with the knowledge to better utilize these tools and advocates for their continuous improvement.

Social media and other online platforms can play a pivotal role in these campaigns. By creating accessible and engaging content that highlights the benefits and practical applications of AI, we can reach a broader audience. Personal stories and testimonials shared online can inspire others and demonstrate the real-world impact of these technologies.

Volunteer programs aimed at supporting AI literacy and deployment in underserved communities can make a significant difference. Volunteers—be they tech enthusiasts, students, or professionals—can provide valuable assistance in teaching people how to use AI tools, setting up devices, and troubleshooting common issues. Such initiatives foster a sense of solidarity and ensure that advancements in AI benefit everyone, not just those with easy access to technology.

A crucial but often overlooked aspect is the involvement of children and young adults with disabilities in STEM fields. Encouraging their participation in tech-related education and activities can spark an interest early on. Programs designed to teach coding, robotics, and AI principles to young individuals with disabilities can empower them to become future innovators who develop solutions not only for themselves but for the broader community.

Another layer of community involvement is legislative advocacy. Engaged communities can work collectively to push for policies and funding that support inclusive AI technologies. This might include campaigning for grants, subsidies for assistive devices, or educational programs that focus on the intersection of disability and technology. Advocacy can serve as a powerful tool to ensure that these technologies not only advance in complexity but also remain accessible and affordable.

Collaboration with local educational institutions can also pave the way for inclusive AI education. Universities and schools can develop specialized courses that focus on the development of assistive technologies, involving students in real-life projects that directly benefit the disability community. By fostering a culture of inclusivity in education, we ensure that the next generation of developers and innovators are well-versed in and committed to creating accessible AI.

Furthermore, we must acknowledge the ethical implications of AI technologies and ensure the community's voice is present in these discussions. Developing ethical guidelines requires input from those who are directly affected by these technologies. Ethical advisory panels and boards that include representatives from the disability community can provide valuable insights into the potential benefits and risks of new developments.

Engagement platforms shouldn't just be limited to feedback and discussions; they should also facilitate co-creation. Inviting community members to be part of the design and development teams allows for the creation of more personalized and effective solutions. These collaborative efforts can lead to innovations that are not only user-friendly but also revolutionary in terms of functionality.

Inclusivity in AI is a continuous journey rather than a destination. It requires constant evaluation, adaptation, and community engagement. Inclusion is more than a checkbox; it's a commitment to ensuring that everyone, irrespective of their abilities, has an equal opportunity to benefit from technological advancements.

Ultimately, the goal is to create an AI ecosystem that is not just inclusive in usage, but also in development and evolution. By embedding community involvement into the very fabric of AI development, we move closer to a world where technology serves as a bridge, not a barrier, enhancing accessibility, independence, and quality of life for people with disabilities.

Chapter 25:
Personal Stories and Testimonials

Throughout this book, we've delved into the multifaceted ways AI can enhance the lives of individuals with disabilities, but perhaps nothing illustrates this more compellingly than personal stories and testimonials. Real-life experiences from people whose lives have been transformed by AI-driven technologies offer a profound look into the impact of these innovations. Whether it's a young student with a learning disability flourishing in a personalized educational environment or an elderly individual regaining independence through advanced mobility aids, these narratives highlight the tangible benefits and emotional uplift that AI technologies can provide. They serve as powerful reminders of why continued investment and ethical development in AI are crucial. These stories aren't just anecdotes; they're testaments to the possibilities that arise when technology meets empathy and innovation meets need. As we aim for an inclusive future, the voices of those directly impacted by these advancements are both inspirational and foundational to our journey.

Real-Life Experiences

Stories of AI's impact on the lives of people with disabilities offer a profound insight into the real-world applications of these technologies. These experiences bring to light not only the capabilities of AI but also its potential to transform lives by breaking down barriers and fostering independence.

Take, for example, Jessica, a young woman with a visual impairment. She discovered a whole new level of autonomy when she began using a sophisticated AI-powered app on her smartphone. The app helps her navigate unfamiliar environments, read text in real-time, and even recognize faces. "It's like having a best friend with me all the time," Jessica says. "I no longer feel hesitant to explore new places or meet new people." Her story exemplifies how AI can augment human capabilities and enrich lives in everyday settings.

Similarly, John, a veteran who lost his ability to walk, shares a story of renewed mobility and a regained sense of purpose. Through advanced robotic exoskeletons and AI-driven physical therapy routines, John has been able to stand and walk with assistance. "The first time I stood up on my own, I felt a surge of emotion that I can't describe," he recalls. "AI didn't just give me my legs back; it gave me a new lease on life." His journey reflects the empowering nature of AI technologies designed to support physical disabilities.

AI isn't just about physical assistance; it also plays a crucial role in social and emotional support. Emma, who has autism, uses an AI-based communication assistant to navigate social interactions more confidently. The assistant provides real-time suggestions, helping her respond appropriately in conversations. "Before this, social gatherings were a nightmare for me," Emma admits. "Now, I actually look forward to parties and meeting new people." Emma's experience underscores the importance of AI in assisting cognitive and social aspects of life.

In the classroom, AI is transforming educational experiences for students like Michael, who has dyslexia. Adaptive learning platforms tailored to his individual needs have made a significant difference. "For the first time, I'm not just keeping up; I'm excelling," he says. The personalized education he receives helps bridge the gap, allowing him to achieve his academic goals. Michael's story illustrates the potential of AI to create inclusive and effective learning environments.

Alice's experience with AI in healthcare is another testament to the technology's life-changing impact. Diagnosed with a chronic illness, she uses an AI-powered health monitoring system that tracks her condition and provides reminders for medication. "Managing my health has become so much easier," Alice states. "I can focus more on living my life rather than being consumed by my illness." The AI system serves as a virtual caregiver, ensuring she remains proactive about her health.

For those with hearing impairments, AI developments like real-time transcription and AI-enhanced hearing aids are making a world of difference. Samuel, who lost his hearing in his 40s, shares that AI-powered hearing aids have reconnected him with his surroundings. The adaptive technology filters out background noise and focuses on relevant sounds, allowing him to engage in conversations effortlessly. "I can hear my grandchildren's laughter again. It's priceless," Samuel shares, beaming with joy.

The impact of AI on the workplace cannot be overstated. Vanessa's story is particularly inspiring; as a web developer with limited mobility, she uses voice-controlled AI tools to write code and manage projects. "I used to worry that my condition would hold me back, but AI has leveled the playing field," she explains. Her ability to contribute effectively at work has boosted her confidence and career prospects.

Then there's Brian, who has struggled with severe anxiety and depression. AI-based mental health apps provide him with immediate access to coping strategies and connect him with mental health professionals. "The AI app on my phone has been a lifeline," Brian confesses. "It's like having a therapist in my pocket." His experience highlights the importance of emotional support systems powered by

AI, offering timely interventions that can make a significant difference in mental health management.

These testimonials are just a few among countless others that demonstrate the transformative power of AI in enhancing the lives of individuals with disabilities. The technology not only offers practical solutions to daily challenges but also instills a sense of empowerment and hope. By illustrating real-life experiences, we see the tangible benefits and the immense potential that AI holds for the future.

Janet, for instance, uses an AI-driven speech recognition device to communicate effectively despite her speech impediment. Before using the device, she felt isolated and struggled to make herself understood. "It was incredibly frustrating," she recalls. "Now, I can participate in meetings and have conversations without feeling self-conscious." Janet's improved communication ability has had a ripple effect, enhancing her social interactions and professional life.

Meanwhile, in another corner of the world, Malik, a student with cerebral palsy, is excelling academically due to AI-enhanced educational tools. Adaptive learning software provides him with the needed accommodations, allowing him to complete assignments and participate in classroom activities. "I used to fall behind, but this technology has allowed me to keep pace with my peers," he shares. Malik's journey is a testament to how AI-powered educational tools can democratize learning and provide equal opportunities for all students.

Older adults like Evelyn also reap the benefits of AI. Living alone, Evelyn relies on an AI home assistant to help with daily activities, from reminding her to take her medications to controlling her home's smart appliances. "It's like having a personal assistant around the clock," she says, emphasizing how AI has brought peace of mind and a greater sense of security to her life. Evelyn's story opens a window into how AI can support independent living for seniors.

In another inspiring account, Thomas, who lost the use of his arms due to an accident, shares how AI-powered voice-controlled devices have restored his ability to write and express himself creatively. "I thought I'd never be able to write again," he confides. "But now, I'm working on my second novel." The technology has allowed him to reclaim his passion, proving that AI can rekindle dreams and aspirations that physical limitations might otherwise stifle.

Even in remote communities, AI is making an impact. Maria, a teacher in a rural area, uses AI tools to enhance her teaching methods and provide personalized instruction to her students with various learning difficulties. "These tools have changed the way I teach," Maria explains. "Now, every child in my class gets the attention they need." Her story highlights the far-reaching potential of AI to bridge educational gaps, regardless of geographical constraints.

The stories shared here are more than just individual triumphs; they represent the collective potential of AI to foster a more inclusive society. The adaptability and scope of AI technologies make it possible to address a wide range of disabilities, thereby improving quality of life for many. These personal testimonies not only inspire but also encourage further innovation and adoption of AI technologies in various sectors.

Moreover, they challenge us to think about the ethical and inclusive design principles necessary to ensure AI systems are accessible to everyone. As we move forward,

Impact of AI on Individual Lives

Stories of how artificial intelligence (AI) has revolutionized the lives of individuals with disabilities are as diverse as the people themselves. When we talk about the impact of AI on individual lives, it becomes a mosaic of personal victories, dramatic improvements in daily living, and newfound opportunities. What ties these stories together is a common thread of enhanced accessibility and independence, which AI makes increasingly possible.

Take, for example, Michael, a visually impaired software developer who has benefited significantly from AI-powered screen readers. Prior to adopting AI tools, he struggled to navigate digital interfaces and relied heavily on memory and pattern recognition to write code. Now, with the help of AI, Michael can interpret complex graphs and charts that were previously inaccessible. His productivity has soared, and he feels more in control of his work environment. AI, in this case, wasn't just a tool; it was a key to unlocking potentials he always knew he had but couldn't fully access.

Or consider Emily, a teenager with cerebral palsy who has found her voice through AI-enhanced communication devices. Traditional communication boards were cumbersome and slow, limiting her interactions to basic needs and simple phrases. Modern AI applications have changed the game by offering predictive text and faster processing times, making conversations with her peers and teachers more fluid and less draining. Emily's parents have noticed a significant improvement in her emotional well-being and social engagement, something that traditional therapies couldn't fully address.

A different, yet equally compelling example, is John, a retiree diagnosed with early-stage Alzheimer's. He relies on AI-driven reminders and cognitive aids to manage his daily routine. These systems learn from his behavior and adapt to his needs, providing prompts that are not just timely but also contextually relevant. tasks that once overwhelmed him are now much more manageable, allowing him to maintain his independence longer than he otherwise could have. John says that these AI tools are like having a "second brain" that never tires and is always ready to assist.

For seniors like Jane, who suffers from severe arthritis, AI-powered robotic assistants have made a monumental difference. Simple tasks such as buttoning a shirt or preparing a meal, once painful and time-consuming chores, are now easier and less stressful. Using sensors and sophisticated AI algorithms, these robotic aides can perform a variety of tasks tailored to individual needs. Jane now has more energy to spend time with her grandchildren and pursue her hobbies, as the burden of daily tasks has significantly diminished.

Meanwhile, AI's impact on educational experiences cannot be understated.

Let's talk about Sam, an elementary school student with dyslexia who once dreaded reading assignments. AI tools designed for personalized learning have transformed his educational experience. Adaptive learning software tailors content to match his reading level and provides interactive feedback, making learning more engaging and less frustrating. These tools have turned his academic struggles into triumphs. Sam's increased confidence has spilled over into other areas of his life, enhancing his overall quality of life.

Another remarkable story comes from Dan, who uses a wheelchair due to a spinal cord injury. Traditional mobility aids were functional but limited in scope, often leaving him feeling dependent on others for day-to-day activities. With the advent of AI-driven mobility solutions, Dan now has access to robotic exoskeletons and AI-powered navigation systems that offer unprecedented levels of autonomy. These innovations have allowed him to explore environments he once thought were off-limits, opening up a world of possibilities and adventures.

AI's contributions aren't confined to physical health and mobility; mental health is another critical area where AI shines. For instance, Maria has been using an AI-powered mental health app to manage her anxiety. The app offers real-time emotional support through chatbots and personalized coping strategies based on her interactions. Unlike traditional therapy sessions that require fixed appointments, Maria can access support whenever she needs it, making her feel less alone in her struggles. The app's ability to adapt to her needs and provide immediate feedback has been a lifeline.

Then there is Alex, a talented musician who lost his hearing at a young age. Advanced AI algorithms have enabled the creation of sophisticated cochlear implants that translate sound waves into electronic signals his brain can interpret. For Alex, this breakthrough means he can continue to compose and enjoy music, a passion that defines his identity. AI has made what seemed impossible possible, bringing sound back into his world in a meaningful way.

It's not just individuals alone; families benefit too.

Susan, a single mother to a child with autism, found herself overwhelmed with the constant demands of caregiving. AI-assisted care management systems track her child's behavior patterns and provide actionable insights, enabling Susan to intervene at the right moments and anticipate her child's needs more effectively. The emotional and psychological relief from constant vigilance has made a significant difference in Susan's quality of life.

AI also has profound implications for workplace accessibility. Take the case of Ahmed, who has cerebral palsy and works as a graphic designer. AI-driven software assists him in creating complex designs with minimal manual input, leveling the playing field and allowing him to compete on equal terms with his colleagues. This has opened up career opportunities that were previously inaccessible, enabling Ahmed to pursue his passion and contribute meaningfully to his field.

For many individuals living with disabilities, the journey of gaining independence and accessibility through AI isn't just about embracing new technology. It's about reclaiming dignity, breaking down barriers, and envisioning possibilities they hadn't dared to dream of. The transformative power of AI can be both tangible and immeasurable, aiding people in ways that affect both their everyday tasks and their broader life goals.

It's a revolution but on a deeply personal scale.

Ultimately, the stories of Michael, Emily, John, Jane, Sam, Dan, Maria, Alex, Susan, and Ahmed are testaments to AI's potential to dramatically enhance individual lives. These narratives underscore that AI is not a distant, impersonal force but a deeply human-focused innovation. By addressing real-world challenges and providing tailored solutions, AI empowers people with disabilities to live fuller, more independent lives. Each story is a chapter in a larger story of progress and empowerment, painting a future where technology and humanity co-create a more inclusive world.

Conclusion

As we conclude this exploration into the transformative power of AI technologies for people with disabilities, it's essential to take a moment to reflect on the extraordinary potential these innovations hold. Artificial intelligence, as we have seen throughout this book, serves not merely as a technological advancement but as a profound tool for empowerment. By bridging gaps in accessibility, providing personalized assistance, and enhancing quality of life, AI is reshaping our world in ways that were once thought to be the realm of science fiction.

AI is more than a collection of algorithms and data; it is a catalyst for change. We've broken down barriers, tackled stereotypes, and created new opportunities for independence and self-sufficiency. From smart assistants that simplify daily routines to adaptive educational tools that cater to individual learning needs, AI technologies have proven to be game-changers. They are not just aiding those with disabilities but are also enriching educational methods, revamping healthcare, and promoting inclusive designs that benefit everyone.

But the journey doesn't stop here. The landscape of AI is evolving rapidly, and its potential applications are growing exponentially. Emerging technologies like neural interfaces, enhanced machine vision, and sophisticated natural language processing are poised to revolutionize accessibility even further. For example, advancements in wearable technology and implantable devices are on the horizon, promising to deliver unprecedented levels of independence for people with disabilities. This continual push for innovation underlines the importance of ongoing research and development.

We must also address the challenges and barriers that come with these technological strides. Ethical considerations are paramount. Ensuring privacy, security, and fair access to AI tools remains a critical focus. It's crucial to integrate comprehensive policies and guidelines that prioritize the rights and dignity of individuals with disabilities. Developers, policymakers, and advocates must work hand-in-hand to create an environment in which technology serves everyone equitably and responsibly.

Realizing the true potential of AI in enhancing the lives of people with disabilities requires collaboration across multiple sectors. Governments, private enterprises, academic institutions, and non-profit organizations all have a role to play. Through partnerships and joint ventures, we can build a robust support system that nurtures innovation while grounded in ethical and inclusive principles.

Let's also not underestimate the importance of user-centered design. Involving people with disabilities in the development process ensures that the products and services tailored for them genuinely meet their needs. Their insights and feedback are invaluable in refining and perfecting assistive technologies. Inclusion in the design and development process promotes empathy, understanding, and, ultimately, better outcomes.

One cannot overlook the impact of personal stories and testimonials. Real-life experiences provide a powerful testament to the efficacy of AI technologies. These narratives put a human face on the data and statistics, reminding us of the real-world implications of our technological advances. They inspire and inform future developments, keeping us grounded in the reality that our ultimate goal is to enhance lives, not just create sophisticated tools.

Additionally, while many innovative solutions have been developed, it's vital to remember that accessibility must be universal. The global perspective shows varied levels of adoption and implementation, highlighting the need for international cooperation and standardization. Socio-economic factors should not be a barrier to access. Efforts must continue to make AI technologies affordable and available worldwide, ensuring no one is left behind.

The path forward is one of promise and responsibility. Future trends and emerging technologies must be watched and harnessed judiciously. It's not just about what's possible; it's about what's beneficial. As AI continues to evolve, so should our approach to integrating it into the lives of people with disabilities. We must remain vigilant and proactive, recognizing that with great power comes great responsibility.

In conclusion, the journey of AI and its impact on individuals with disabilities is a testament to human ingenuity and compassion. It paints a picture of a future where barriers are dismantled, and possibilities are boundless. This is a call to action for all stakeholders to continue pushing the envelope, breaking new ground, and ensuring that the AI ecosystem we create is inclusive, ethical, and supportive of all its users.

The future indeed looks bright. As we've seen, technological advancements in AI are unlocking new levels of accessibility and independence. What's imperative now is that we foster a culture of continuous improvement and inclusivity. By doing so, we ensure that AI lives up to its potential as a powerful ally in creating a world where disabilities are not barriers but simply different ways of experiencing life.

Let's commit to this vision. Together, we can build a future where technology and human potential intersect to create a society where everyone, regardless of their abilities, has the opportunity to thrive and contribute. It's not just an aspiration; it's an achievable reality, one innovation, and collaboration at a time.

Appendix A:
Appendix

The purpose of this appendix is to provide additional information and resources that complement the main content of our book on how artificial intelligence can empower individuals with disabilities. This section is designed to serve as a quick-reference guide, offering further insights, clarifications, and practical tools that will help you better understand and leverage the transformative power of AI in improving accessibility, independence, and quality of life.

Supplementary Resources

Here, you'll find a curated list of supplemental resources to deepen your understanding of AI and its applications for people with disabilities. These include:

- Books and Articles: Additional reading materials that delve deeper into topics discussed in the book.

- Websites and Online Platforms: Online resources and communities dedicated to AI and accessibility.

- Educational Courses: Courses and training programs that offer more formal education on AI technologies and their applications.

Glossary of Terms

To make the complex jargon more accessible, we've included a glossary of important terms related to AI and disability. This glossary aims to clarify technical language and ensure that all readers, regardless of their background, can grasp the concepts discussed.

Case Study Summaries

For quick reference, this section offers condensed versions of the case studies detailed throughout the book. These summaries highlight key points, successes, challenges, and lessons learned from each case study, providing at-a-glance information that can be useful for presentations, discussions, and further exploration.

Toolkits and Checklists

We've prepared some practical toolkits and checklists to assist you in applying the knowledge gained from the book. These include:

- Accessibility Checklists: Guidelines for evaluating and improving the accessibility of various technologies and environments.
- Development Toolkits: Resources for developers to create inclusive AI applications.
- Implementation Guides: Step-by-step instructions for integrating AI solutions into existing systems and practices.

Contact Information

If you have any questions, need further clarification, or wish to share your experiences and feedback, we've included contact information for reaching out to the authors, contributing experts, and support organizations.

Additional Case Studies and Testimonials

Beyond the principal case studies and personal stories featured in the book, this section provides extra examples and testimonials. These additional narratives showcase the broader impact of AI on various aspects of life for people with disabilities, offering more real-world insights and inspiration.

Research and Development Projects

This part of the appendix lists ongoing and upcoming research projects focused on AI and accessibility. By highlighting these initiatives, we aim to connect interested readers with cutting-edge developments and opportunities for collaboration, further study, and potential funding.

We hope this appendix serves as a valuable extension of the book, helping you to maximize the benefits of AI for enhancing accessibility and empowerment. Whether you're an educator, healthcare professional, tech enthusiast, or an individual with disabilities, these resources are here to support your journey towards a more inclusive future.

Glossary of Terms

This glossary aims to provide clear definitions of terms that are essential for understanding the intersection of artificial intelligence and disability. Familiarizing yourself with these terms will enhance your comprehension of the material covered in this book.

Accessibility: The design and creation of environments, products, and services that are usable by all people, including those with disabilities, without the need for adaptation.

Artificial Intelligence (AI): The simulation of human intelligence processes by machines, especially computer systems. This includes learning, reasoning, and self-correction.

Assistive Technology: Devices or systems that help people with disabilities perform tasks that might otherwise be difficult or impossible.

Augmentative and Alternative Communication (AAC): Methods of communication used to supplement or replace speech for individuals with impairments in expressive communication.

Autonomous Systems: Systems that can perform tasks without human intervention, often leveraging AI to make decisions in real-time.

Cognitive Disabilities: Disabilities that affect learning, memory, problem-solving, and attention. Examples include intellectual disabilities, autism, and traumatic brain injury.

Deep Learning: A subset of machine learning involving neural networks with many layers ('deep' networks) that can learn from vast amounts of data.

Emotional Wellbeing Tools: AI-based tools designed to monitor and improve mental health by providing support, such as chatbots or mood tracking apps.

Inclusive Design: Design practice that ensures products, environments, and experiences are accessible to, and usable by, as many people as possible, including those with disabilities.

Machine Learning: A branch of AI that enables computers to learn from and make predictions based on data. It involves algorithms that improve their performance as they are exposed to more data over time.

Natural Language Processing (NLP): A field of AI focused on the interaction between computers and humans through natural language. It enables computers to understand, interpret, and generate human language.

Personalized Learning Environments: Educational settings or tools tailored to meet the individual needs and preferences of learners, often using AI to adapt content and methods.

Robotic Assistance: The use of robots to support individuals with physical disabilities in performing daily tasks, increasing their independence and quality of life.

Smart Assistants: AI-powered devices or applications like virtual assistants (e.g., Siri, Alexa) that help users perform tasks, retrieve information, and manage daily activities.

Speech Recognition: The ability of a machine or program to identify words and phrases in spoken language and convert them into machine-readable format.

Telehealth: The use of electronic information and telecommunications technologies to support long-distance clinical healthcare, patient and professional health-related education, and public health.

Universal Design: The concept of designing products and environments in a way that they can be used by all people, to the greatest extent possible, without the need for adaptation or specialized design.

Virtual Assistants: AI systems designed to help users by performing tasks or services, often through voice interactions. Examples include Google Assistant, Amazon Alexa, and Apple Siri.

Additional Resources

Expanding your knowledge and resources beyond the glossary of terms is crucial in understanding how to harness AI for the benefit of people with disabilities. There are numerous additional resources that can enrich your insights, including specialized publications, online platforms, community organizations, and more. Below we provide a range of resources that can help you delve deeper into the topics discussed in this glossary and the broader context of AI and disability inclusion.

First, scholarly articles and journals are invaluable for staying current with the latest research. Academic databases like PubMed, IEEE Xplore, and Google Scholar offer extensive collections of peer-reviewed papers that explore the multifaceted relationship between AI and disability. Journals specifically focused on assistive technology, rehabilitation engineering, and human-computer interaction often feature groundbreaking studies and are essential reading for anyone looking to impact this field.

Industry reports and whitepapers issued by tech companies, research institutions, and advocacy organizations can provide pragmatic insights into current trends and future directions. Reports from organizations like the World Health Organization (WHO), the National Institute on Disability, Independent Living, and Rehabilitation Research (NIDILRR), and the AI Now Institute often include data-driven analyses and policy recommendations, highlighting practical aspects of implementing accessible AI solutions.

Online courses and certification programs can offer structured learning for those new to the field or looking to expand their expertise. Platforms like Coursera, edX, and Udacity have courses focusing on AI, machine learning, and their applications in accessibility and assistive technologies. These courses often feature modules led by experts in the field, making them an excellent way to gain both theoretical knowledge and practical skills.

Community forums and online discussion groups can be an excellent way to stay updated and get advice from both experts and peers. Websites like Reddit, Stack Overflow, and specialized forums for disability advocacy and AI development can provide real-time answers to specific questions and serve as a sounding board for new ideas and collaborative projects. Engaging in these communities not only helps in solving technical challenges but also offers emotional support and a sense of belonging.

For developers and tech enthusiasts, open-source repositories on platforms like GitHub can be invaluable. These repositories often host projects related to AI and accessibility, providing code, documentation, and collaboration opportunities. By contributing to or leveraging existing projects, developers can play an active role in accelerating innovation in this essential field.

Moreover, think tanks and advocacy groups dedicated to disability rights and technology are pivotal resources. Organizations such as the

Global Initiative for Inclusive ICTs (G3ict), the Center for Democracy & Technology (CDT), and the American Association of People with Disabilities (AAPD) produce extensive research and advocacy materials. These resources often include toolkits, policy papers, and guidelines that can help stakeholders navigate the complexities of developing and deploying inclusive AI.

Conferences and workshops also offer immersive learning experiences. Events like the International Conference on Computers Helping People with Special Needs (ICCHP), the annual CSUN Assistive Technology Conference, and the AI for Good Global Summit often feature presentations and sessions led by pioneers in the fields of AI and accessibility. Attending such events can provide valuable opportunities for networking, hands-on demonstrations, and deep dives into specialized topics.

In addition to formal learning and networking opportunities, social media can be a resourceful tool for real-time updates and community engagement. Following hashtags like #A11y (shorthand for "accessibility"), #Tech4Good, and #DisabilityInclusion on platforms such as Twitter and LinkedIn can keep you informed about the latest discussions, events, and innovations in the space.

Finally, books and comprehensive guides can serve as foundational texts and reference materials. Works such as "Artificial Intelligence: A Guide for Thinking Humans" by Melanie Mitchell and "Inclusive Design for a Digital World" by Regine M. Gilbert provide in-depth explorations into AI concepts and their ethical implications in the real world. Reading books from diverse authors and perspectives ensures a well-rounded understanding and can often provide inspiration for implementing your own projects.

It's also beneficial to connect with local and international disability organizations. These groups not only offer support services but often engage in advocacy, research, and partnerships focusing on the development of accessible technologies. Organizations like Disability Rights International and the International Disability Alliance can provide connections and resources on a global scale.

Podcasts and webinars can be an excellent way to consume information on the go. Shows like "AI Alignment," "Raising the Bar on Accessibility," and various tech industry webinars frequently cover relevant topics from both a technical and user-experience perspective. These mediums allow you to hear directly from experts and thought leaders, offering practical advice and the latest insights.

Lastly, government agencies and public institutions often provide grants, funding opportunities, and detailed guidelines for developing accessible technologies. Agencies such as the National Science Foundation (NSF) and the European Union's Horizon 2020 program frequently support research and innovation in this space. Furthermore, examining the accessibility standards set by bodies such as the Web Content Accessibility Guidelines (WCAG) and the Americans with Disabilities Act (ADA) can help ensure compliance and promote best practices.

By leveraging these additional resources, you can better navigate the rapidly evolving landscape of AI and its applications in supporting people with disabilities. Whether you're an educator, healthcare professional, tech enthusiast, or someone with a disability, these tools and platforms can empower you to contribute effectively to creating a more inclusive and accessible world.

Made in the USA
Columbia, SC
01 March 2025